Beverage Quality and Safety

Beverage Quality and Safety

Edited by
Tammy Foster
and
Purnendu C. Vasavada

INSTITUTE OF FOOD TECHNOLOGISTS

CRC PRESS

Boca Raton   London   New York   Washington, D.C.

Library of Congress Cataloging-in-Publication Data

Beverage quality and safety / edited by Tammy Foster and Purnendu C. Vasavada.
        p. cm.
    Includes bibliographical references and index.
    ISBN 0-58716-011-0 (alk. paper)
    1. Beverages—Quality control. 2. Beverage industry—Quality control. I. Foster, Tammy.
II. Vasavada, Purnendu C.

TP511.B48 2003
663'.6'0685—dc21   2003046136

This book contains information obtained from authentic and highly regarded sources. Reprinted material is quoted with permission, and sources are indicated. A wide variety of references are listed. Reasonable efforts have been made to publish reliable data and information, but the author and the publisher cannot assume responsibility for the validity of all materials or for the consequences of their use.

Neither this book nor any part may be reproduced or transmitted in any form or by any means, electronic or mechanical, including photocopying, microfilming, and recording, or by any information storage or retrieval system, without prior permission in writing from the publisher.

All rights reserved. Authorization to photocopy items for internal or personal use, or the personal or internal use of specific clients, may be granted by CRC Press LLC, provided that $1.50 per page photocopied is paid directly to Copyright Clearance Center, 222 Rosewood Drive, Danvers, MA 01923 USA. The fee code for users of the Transactional Reporting Service is ISBN 0-58716-011-0/03/$0.00+$1.50. The fee is subject to change without notice. For organizations that have been granted a photocopy license by the CCC, a separate system of payment has been arranged.

The consent of CRC Press LLC does not extend to copying for general distribution, for promotion, for creating new works, or for resale. Specific permission must be obtained in writing from CRC Press LLC for such copying.

Direct all inquiries to CRC Press LLC, 2000 N.W. Corporate Blvd., Boca Raton, Florida 33431.

Trademark Notice: Product or corporate names may be trademarks or registered trademarks, and are used only for identification and explanation, without intent to infringe.

Visit the CRC Press Web site at www.crcpress.com

© 2003 by CRC Press LLC

No claim to original U.S. Government works
International Standard Book Number 0-58716-011-0
Library of Congress Card Number 2003046136
Printed in the United States of America 1 2 3 4 5 6 7 8 9 0
Printed on acid-free paper

Foreword

As an industry professional, I have always found the Institute of Food Technologists (IFT) to be a valuable educational resource. This book is a result of a workshop entitled Emerging Beverage Technology, in which many of my colleagues presented on a variety of topics. As I look back on what was "emerging" then, I see how these issues have surfaced for beverage manufacturers. Both basic and cutting-edge issues are addressed in this book. This publication covers the basics of plant sanitation, as presented by Martha Hudak-Roos and Bruce Ferree. It goes into depth on Good Agricultural Practices to ensure safe juice, as discussed by Richard Stier and Nancy Nagle. Donald Kautter, who helped develop the Food and Drug Administration's Juice Hazard Analysis and Critical Control Point (HACCP) regulation, speaks directly to the final rule. Emerging issues, such as the roles of genetically modified organisms (GMOs), nutraceuticals, and alternative technologies, are presented by Susan Harlander, Dennis Gordon, Kiyoko Kubomura, and Purnendu Vasavada, respectively.

In order to stay competitive, manufacturers must forever improve their technology, products, and processes. It is not enough to maintain the status quo, or your competitor will suddenly overtake you. Beyond competition, there are always new food safety concerns in the beverage world and new technologies to be explored. As much as consumers want a new and exciting beverage, they never want to worry about its safety. In the quest to satisfy consumers' thirst for new and interesting beverages, technology is key. Academia, industry, and scientific organizations will need to continue to work together to meet consumer expectations. New beverage technology and the opportunity it presents are expanding. The role of innovation will continue to drive the juice and beverage markets and in the end drive consumer loyalty. This publication is only one step in the ongoing process of continuous improvement.

Linda Frelka
Vice President
Odwalla, Inc.
Half Moon Bay, California

Foreword

Beverage Quality and Safety is based on information presented in a program held at the Annual Meeting of the Institute of Food Technologists (IFT). It is compiled from the extensive knowledge of a team of experienced food industry experts, whose expertise is based on many years of direct involvement with the food and beverage industries. Their qualifications are described elsewhere, but their collective dedication in sharing their knowledge with others in the industry has made it possible for the Institute of Food Technologists' Continuing Education Committee not only to present the information provided for this book to readers everywhere, but also to present it as oral educational programs to IFT members and nonmembers. IFT is dedicated to providing the latest technical information relating to food processing, and its Professional Development Department coordinates this effort throughout the year. Topics selected by IFT for presentation and publication are peer reviewed for maximum interest by different segments of the food industry.

The beverage market continues to grow, despite recent setbacks in the world economy. New technology in processing and packaging continues to please consumers with the introduction of new beverage products. We hope this book will act as a reference for researchers, processors, marketers, and consumers. IFT sincerely thanks all of the contributors, and especially the editors, Tammy Foster and Purnendu Vasavada, for their expertise and effort.

Dean D. Duxbury
Director of Professional Development
Institute of Food Technologists
Chicago, Illinois

Preface

The fruit juice, soft drink, and beverage industry has experienced rapid growth in recent years. While traditional drinks and beverages have maintained consumer interest, new, innovative, value-added products, including exotic juice and beverage blends, energy drinks, sports drinks, ready-to-drink teas and coffees, bottled water, and beverages containing nutraceuticals, botanicals, and herbal ingredients have generated much excitement in the beverage sector. The global market for functional foods, estimated to be over $35 billion, is expected to reach 5% of the total world food expenditure in the near future. Beverages constituted a significant proportion (33 to 73%) of various health-promoting new products or product lines introduced in the U.S. in 2000. According to a recent industry report, the U.S. functional beverage market generated revenues of $4.7 billion in 2000 and is predicted to exceed $12 billion by 2007. Another industry report indicated that refrigerated juices, nectars, juice blends, cocktail drinks, and refrigerated teas generated over $3.5 billion and $105 million, respectively, in sales in 2002.

In recognition of the significance of the juice and beverage sector in the food industry, the Institute of Food Technologists (IFT) developed and offered a short course, Beverage Technologies and Regulatory Outlook, as a part of the IFT Continuing Education Program prior to the IFT annual meeting in 2001. The short course was designed to offer information on the latest beverage industry trends and developments relating to products, processing, and packaging technologies and to provide an update on regulatory issues such as federal Hazard Analysis and Critical Control Point (HACCP) regulations and Codex Alimentarius Commission activities related to fruit juice. From discussions with the IFT Continuing Education Committee (IFT-CEC) and industry colleagues, it was felt that a publication providing discussion of the industry and regulatory trends as well as the quality and safety of fruit juice and beverages would be useful. This book contains chapters based on many of the presentations at the short course. It is not intended as a comprehensive review of the details of recent research on the topic of fruit juice and beverage technology. Rather, it is designed to provide an applied, "practitioner's" viewpoint on the fruit juice and beverage industry from "grove to glass."

The book opens with a chapter on minimizing contamination in the production sector followed by a discussion of the role of genetically modified organisms (GMOs) in beverage production. The role of nutraceuticals and functional food applications in beverage production is discussed in Chapter 3. The production and processing of organic fruit, juice, and beverages are detailed in Chapter 9.

The processing and packaging of juices and beverages are discussed in Chapters 4, 9, and 10, and cleaning and sanitation of beverage plants are discussed in Chapter 8. The microbiological aspects of fruit juices and beverages, particularly the importance of microorganisms in spoilage and safety of fruit juice, are discussed in

Chapters 4 and 5. Traditionally, pathogenic organisms were not a major cause for concern in fruit juices and fruit beverages. However, reports of foodborne illness outbreaks, consumer illness, and recalls associated with fruit, fruit juice, and juice products during the past decade have led to a recognition of emerging pathogens as a major threat to the safety of fruit juice and beverages. In the wake of the food safety concerns, the U.S. Food and Drug Administration (FDA) has issued guidance to minimize microbial food safety hazards in fresh and minimally processed fruits and vegetables, required a warning label on any unpasteurized juices, and mandated implementation of the Hazard Analysis Critical Control Point (HACCP) system designed to ensure safety of fruit juice and juice products. Chapters 5, 6, and 7 provide detailed discussions of the design and implementation of HACCP in the juice and beverage industry.

The IFT short course featured a presentation on the Codex activity regarding fruit juice and vegetable juice standards by the FDA representative serving on the U.S. delegation to the Ad Hoc Intergovernmental Task Force on Fruit and Vegetable Juices. We would have liked to include a chapter on the Codex activities dealing with the fruit juice and vegetable juice standards. However, the Codex fruit juice and vegetable juice standards have not been finalized and are being currently debated by the Codex Ad-Hoc Intergovernmental Task Force on Fruit and Vegetable Juices. Detailed reports of recent meetings of the ad-hoc commission are available on the Internet at the U.S. Codex Web site.

We are grateful to all the contributors for providing manuscripts and to Linda Frelka, vice president, Odwalla, Inc., and Dean Duxbury, the IFT director of professional development, for writing Forewords for this book. We would also like to thank Dean Duxbury and the IFT-CEC staff for their encouragement and support. Finally, we would like to thank Eleanor Riemer and Erika Dery of CRC Press for their patience and valuable assistance in the production of this book. The contributors, who are specialists well known in their fields, and the editors have the best intentions and efforts in producing the book and hope that, despite any shortcomings, it will be a useful source of information for professionals in food industry.

Tammy Foster
Purnendu C. Vasavada

About the Editors

Tammy Foster is food safety manager for Tropicana Products, Inc., in Bradenton, Florida. She has held various positions in food microbiology, safety, and quality assurance and is currently responsible for standardizing sanitation programs/systems for Tropicana worldwide, reviewing new equipment and new processes for sanitary design, reviewing and ensuring that Hazard Analysis and Critical Control Point (HACCP) plans are in compliance with federal regulations, and monitoring water quality within all manufacturing facilities. She is a member of the American Society of Quality, the Institute of Food Technologists (IFT), and the International Association for Food Protection (IAFP) and has served as a member and chair of the IFT Continuing Education Committee. Ms. Foster received a B.S. degree in microbiology from South Dakota State University.

Purnendu C. Vasavada is professor of food science at the University of Wisconsin–River Falls and food safety and microbiology specialist with the University of Wisconsin (UW) Extension. He has developed and taught undergraduate courses in food science and technology and has been an invited participant in international conferences, workshops, and symposia dealing with rapid methods and automation in microbiology, food safety and microbiology, food quality assurance, HACCP and TQM (Total Quality Management), and food science education in the U.S., Canada, the U.K., Ireland, Mexico, Australia, New Zealand, Singapore, Malaysia, Argentina, Chile, Brazil, Hungary, Norway, Sweden, and Finland. He has organized the UW River Falls International Food Microbiology Symposium and Rapid Methods in Food Microbiology Workshop for the past 22 years. Dr. Vasavada is author or coauthor of more than 70 publications, including technical abstracts, research papers, book chapters, and articles in professional and trade publications. A fellow of the American Academy of Microbiology, Dr. Vasavada is the recipient of the Joseph Mityas Laboratorian of the Year Award (1987) from the Wisconsin Laboratory Association, the Educator award from the International Association of Milk, Food, and Environmental Sanitarians (IAMFES; 1997), the Sanitarian of the Year award from the Wisconsin Association of Milk and Food Sanitarians (1998), and the Chairman's Award from Minnesota IFT (1998). He is a member of IFT and the International Association for Food Protection and has served as a member and chair of the IFT Continuing Education Committee. He received B.Sc. and M.Sc. degrees in microbiology in India, an M.S. in microbiology from the University of Southwestern Louisiana in Lafayette, and a Ph.D. in food science and dairy manufacturing from the University of Georgia in Athens.

Contributors

Paul L. Dawson
Clemson University
Clemson, South Carolina

Bruce Ferree
Technical Food Information
  Spectrum, Inc.
Lodi, California

Tammy Foster
Tropicana Products, Inc.
Bradenton, Florida

Dennis T. Gordon
North Dakota State University
Fargo, North Dakota

Susan Harlander
BIOrational Consultants, Inc.
New Brighton, Minnesota

Martha Hudak-Roos
Technical Food Information
  Spectrum, Inc
League City, Texas

Donald A. Kautter, Jr.
U.S. Food & Drug Administration
Washington, D.C.

Todd Konietzko
Schwan's Sales Enterprises
Marshall, Minnesota

Kiyoko Kubomura
Kubomura Food Advisory Consultants
Tokyo, Japan

Nancy E. Nagle
Nagle Resources
Pleasanton, California

Richard F. Stier
Consulting Food Scientists
Sonoma, California

Susan Ten Eyck
California Certified Organic Farmers
Santa Cruz, California

Purnendu C. Vasavada
University of Wisconsin
River Falls, Wisconsin

Contents

Chapter 1 Ensuring Safety in Juices and Juice Products: Good
Agricultural Practices 1
Richard F. Stier and Nancy E. Nagle

Chapter 2 The Role of Genetically Modified Organisms (GMOs)
in Beverage Production 9
Susan Harlander

Chapter 3 Beverages as Delivery Systems for Nutraceuticals 15
Dennis T. Gordon and Kiyoko Kubomura

Chapter 4 Alternative Processing Technologies for the Control
of Spoilage Bacteria in Fruit Juices and Beverages 73
Purnendu C. Vasavada

Chapter 5 Microbiology of Fruit Juice and Beverages 95
Purnendu C. Vasavada

Chapter 6 U.S. Food and Drug Administration:
Juice HACCP — The Final Rule 125
Donald A. Kautter, Jr.

Chapter 7 HACCP:
An Applied Approach 157
Todd Konietzko

Chapter 8 Essential Elements of Sanitation in the Beverage Industry 175
Martha Hudak-Roos and Bruce Ferree

Chapter 9 Juice Processing — The Organic Alternative 193
Susan Ten Eyck

Chapter 10 Active Packaging for Beverages 205
Paul L. Dawson

Index 219

1 Ensuring Safety in Juices and Juice Products: Good Agricultural Practices

Richard F. Stier and Nancy E. Nagle

CONTENTS

Introduction 1
Evolution of GAPs 2
Microbiological and Chemical Safety 3
Certification 5
The Proactive Approach Is Good Business 6
Summary 7
References 7

INTRODUCTION

The emphasis on food safety has led to the adoption of the HACCP (Hazard Analysis and Critical Control Points) system by food processors throughout the world. Adoption has been both voluntary and mandatory, as food regulatory agencies have moved to mandate the system for different products. In the United States, HACCP has been mandated for the juice processing industry. Codex Alimentarius, the body aimed at developing guidelines for international trade, has also adopted HACCP as part of its Code of Food Hygiene. In fact, if you talk to delegates to the Codex Committee on Food Hygiene, you will learn that HACCP literally "sailed" through the Committee. Adoption of the system took only a few years, which is incredible when one understands that Codex is an organization in which change may take decades.

HACCP is a system that was developed to ensure the safety of processed foods, so this leaves a great deal of the food supply "uncovered." Why do we say "uncovered"? We say it because HACCP is a system in which a food processor identifies potential hazards and builds "controls" into the process to eliminate, reduce, or control each hazard. With fresh produce, this is not realistic, as it is literally impossible to eliminate or control all potential hazards. Processes designed to destroy or control most pathogens would change fresh products so that they would no longer be fresh. Understanding this, representatives from industry, government, and academia took steps to remedy this deficiency. They developed what are now called Good Agricultural Practices or GAPs. The GAPs are a logical extension of HACCP into the fresh produce industry. They utilize HACCP principles and prerequisite programs to reduce the potential for product contamination and thereby ensure safety. Recent activities at the International Organization for Standardization (ISO) further underscore the importance of food safety. ISO is in the process of developing food safety standards that address both HACCP and Good Agricultural Practices.[1]

What is interesting is that many food processors who are buying produce are now mandating that the materials be purchased from growers who operate under GAPs. This applies even when the fresh products are being further processed. These companies operate under the theory that the application of GAPs will help to ensure the safety of their products, and thus protect their customers, business, and reputation.

EVOLUTION OF GAPs

Good Agricultural Practices continue to evolve throughout the world. In the United States, the Western Growers Association, the International Fresh Cut Produce Association, the government, and industry have been and remain active in their efforts to develop training tools and other documentation to ensure that growers produce foods that are free from foodborne hazards. The *Guide to Minimize Microbial Food Safety Hazards for Fresh Fruits and Vegetables*,[2] released by the U.S. Food and Drug Administration (FDA) on October 26, 1998, addresses microbiological food safety. Chemical hazards are addressed in other documents. In Europe, industry and government are following a similar path. The EUREGAP certification protocols[3] define "best practices" for global production of horticultural products. The key word here is "global." As denizens of First World nations continue to demand fresh foods year round, they must turn more and more to less developed nations to supply these products. But the demands do not stop at the foodstuffs themselves. These same people (and their governments) also demand that the produce that crosses international boundaries be safe and wholesome. The key to ensuring the safety of produce that enters the world market is the development and implementation of Good Agricultural Practices. As an example, if a grower in Central Africa wished to market fresh green beans into Europe, that grower would need to adopt GAPs. Along these same lines, it would not be unreasonable for buyers of juice concentrates or purees to mandate that their vendors ask their suppliers of fruit to adopt Good Agricultural Practices, even if the products are going to be pasteurized prior to sale.

The GAP protocols are science-based systems and are designed to ensure to a high degree of confidence that produce is safe. As one reads over the guidelines that have been developed, it is easy to see that what people once called "common sense" also characterizes these guidelines. The common-sense practices have simply been codified. Adoption of these practices, which may also be applied to fruits and vegetables destined for processing or those used as ingredients, is seen as a burden in many producing countries in the Third World. There are many in these nations who also perceive GAPs to be unfair barriers to trade that have been "foisted" upon them by the more affluent nations. This perception is way off the mark. The adoption of GAPs will help producers in developing countries not only to build their businesses but also to protect those businesses once they are established. One only needs to look at Nicaragua and its raspberries to see how failure to adopt procedures has hurt a whole nation. But the development of food safety programs in these nations is not something that will be accomplished quickly or easily. Cultural, regulatory, and educational constraints can hinder such growth.[4] If buyers for juice processors are going to look "far and wide" for unique concentrates or purees, they should also be willing to work with vendors to help them upgrade programs from "farm to fork."

Recent efforts in Belgium provide an excellent example of how adoption of GAPs can help build and maintain businesses. To ensure that the nation is able to meet the quality and safety demands of its customers, the Belgian Federation of Vegetable Trading and Processing Companies has established a Quality and Food Safety System.[5] This system addresses the whole food chain (farmers, contractors, traders, processors, and distributors) and integrates existing recordkeeping programs that have been implemented as part of HACCP or ISO 9000. The Centrum voor Kwaliteitscontrole (CKC), a nonprofit center, was created to monitor the system. The CKC seeks accreditation from the Belgian Food Safety Agency and EUREGAP accepted in the future.

MICROBIOLOGICAL AND CHEMICAL SAFETY

Microbiological food safety was the driving force behind the development of Good Agricultural Practices in the United States. A review of past literature reveals that an increasing number of foodborne outbreaks has been associated with fresh produce in recent years. In some of these, such as the tragic event involving radish sprouts in Sasaki, Japan, deaths occurred. Juices and juice products have also been implicated in food poisoning outbreaks (Table 1.1).[7] Unprocessed juices have been the source in almost every instance. A similar

TABLE 1.1
Foodborne Illnesses Attributed to Juice Products

Product	Year	Microorganism
Apple cider	1922	*Salmonella typhimurium*
Apple cider	1975	*S. typhimurium*
Apple cider	1982	*Escherichia coli* O157:H7
Apple cider	1991	*E. coli* O157:H7
Orange juice	1995	*S. hartford*
Apple juice	1996	*E. coli* O157:H7

Source: From Stier, R.F., GMPs and HACCP for Beverages, short course sponsored by the Institute of Food Technologists, 1998.

review of the literature in 5 or 10 years should help document whether the implementation of GAPs has made a difference. Since some processors still market fresh juices, it would make sense that these processors make an effort to mandate that their suppliers of fresh fruits or vegetables adopt GAPs. For example, the guideline that says apples used in the manufacture of fresh cider or apple juice be harvested from the tree and not picked off the ground is one such practice.

Ensuring microbiological safety of fresh fruits and vegetables, whether destined for the fresh market or for further processing, is a task that requires a company-wide commitment, but one cannot ignore potential chemical hazards, either. In fact, potential chemical contamination from pesticides may be an even greater concern when buying produce or processed juice concentrates or purees from Third World nations. The amount of pesticide on a product may not be enough to cause illness, but it can surely result in a product being denied entry to an importing country or exit from an exporting nation. For example, many nations have established export authorities whose main mission is to test products destined for export. Without a certificate from this state-run laboratory, the product cannot move forward. This places a burden on growers, and, as has been emphasized time and again, does little to ensure food safety. Safety is best ensured by development, implementation, and adherence to a well-designed control program, rather than by what amounts to random sampling. This mentality was underscored at the Codex Coordinating Committee Meeting in Cairo in January 2001. The delegates initiated a movement to develop sampling procedures and guidelines to ensure food safety. After a rather lengthy discussion, Dr. Alan Randall from the Food and Agriculture Association in Rome took over the floor and explained that the Codex Committee on Food Hygiene has adopted HACCP as the best tool for ensuring food safety and that testing was not the way to go. The bottom line is that there are inherent biases throughout the world when it comes to a systematic and proactive approach to food safety employing HACCP or Good Agricultural Practices.

As noted earlier, there is a "push" the world over to ensure food safety. The United Fresh Fruit and Vegetable Association has a working group that has been working on a Food Safety Questionnaire for Fresh Fruits and Vegetables.[6] This document should be complete by the time that this book is published. The questionnaire uses the FDA's "Guide" as the basis for designing questions but incorporates questions that emphasize chemical safety as well. The stated objective of the questionnaire is to "assess how or if food safety issues are addressed in the production and distribution of fruits and vegetables." The document emphasizes that there are no right or wrong answers. It has been designed to be user friendly and help the grower or packer better understand potential risks and where more work may be needed. It is very similar to the EUREGAP Protocol for Fresh Fruits and Vegetables.[3] The principal difference is that EUREGAP Protocols are mandatory rules that must be followed if an operation wishes to be certified. Certification issues will be addressed at greater length later.

The human element is, perhaps, the most difficult of all to control. Growers can provide proper facilities, conduct what they feel are adequate worker education programs, and pay their workers a fair wage, but the bottom line is that the large majority of field and packing house workers are at the lower ends of the economic and education spectrums. All too often, they see the work as simply a job and are not aware of (or may not care about) the consequences of their actions. This is why worker education programs must not only address basic hygiene issues, but also be relevant to the employees' work and life. For example, consultants have been successful in teaching food safety and hygiene to the predominantly female agricultural workforce in Egypt. They found that the women were eager to learn methods that would help them keep their own families safe. This is definitely an issue with regard to developing food safety programs in developing nations.[4]

CERTIFICATION

Europeans place a greater emphasis on certification than North Americans do. ISO, HACCP, and GAP certification are much more prominent on that side of the Atlantic. The EUREGAP protocols are the guidelines that growers, distributors, and packing houses must meet if they wish to be certified and to sell their products into certain markets or to established buyers. The EUREGAP protocols include both required and encouraged (recommended) practices. They do not specify exactly how the requirements are to be achieved, however. The producer therefore has a certain leeway in meeting the goals.

EUREGAP is in the process of evaluating certifying agencies from around the world. The vast majority of these are European firms, but the United States is represented by companies such as Scientific Certification Systems (SCS) of Oakland, California and Primus Labs of Santa Maria, California. Both of these operations have actively worked with growers and packers in California and Mexico and have assisted in the development of programs to enhance the safety of produce.

Certification has its pros and cons. Obviously, any company that has made the effort to be certified has a certain amount of discipline. It has met the requirements of the certifying agency, which for GAPs includes development of programs and documentation of those activities. Areas where programs need to be in place include site history; fertilizer usage; irrigation; chemical use and storage; crop protection; harvesting; postharvest handling and treatments; waste; worker health, safety, and education; and environmental issues. The ultimate goal is consumer health and therefore, customer satisfaction. On the other hand, certifying agencies and the companies that they certify must avoid falling into the trap of thinking that Good Agricultural Practices and their maintenance are exercises in recordkeeping. GAPs, like HACCP, are a system to ensure the production of safe foods. If the program goes from a quality/safety system to one where the documents take precedence, the program will be compromised. This is precisely what has happened with ISO 9000, and it is one of the reasons that ISO 9000 2000 has incorporated customer satisfaction into the new programs.

THE PROACTIVE APPROACH IS GOOD BUSINESS

In certain areas, certification will be mandatory for people to do business. Certification is also a means whereby growers or packers can demonstrate their commitment to the production and distribution of safe foods. The certificate then becomes a marketing tool that allows them to enter markets previously out of reach.

Adoption of Good Agricultural Practices has another benefit that all persons involved in the food business need to understand. The law requires that the foods you distribute be safe and wholesome. It is good business to do all in your power to achieve this goal. Failure to adopt and follow what are acknowledged as "best practices" can have significant adverse economic consequences in the event that a food safety problem occurs. Look at two of the more high-profile outbreaks over the past few years: Sara Lee's cooked meat products and Odwalla's juice. Products manufactured by both companies were implicated in outbreaks of foodborne illness, and because the companies failed to follow best practices (due diligence), their penalties were much greater. The potential costs of failing to "do it right" can be high.

SUMMARY

Good Agricultural Practices (GAPs) are a means to help ensure the safety of fresh fruits and vegetables. Traditionally, they are usually applied to produce destined for the fresh market, but because of the emphasis on enhanced safety, more and more buyers of fruits and vegetables for further processing are asking that the raw materials be produced using the principles of Good Agricultural Practices. This is especially true in the juice industry, since there are still many "fresh" juices on the market.

REFERENCES

1. Surak, J., personal communication, 2002.
2. U.S. Department of Health and Human Services, Food and Drug Administration, *Guide to Minimize Microbial Food Safety Hazards for Fresh Fruits and Vegetables,* October 26, 1998.
3. EUREGAP Protocol for Fresh Fruits and Vegetables, 2001.
4. Stier, R.F., Ahmed, M.S., and Weinstein, H., Constraints to HACCP implementation in developing nations, *Food Safety Magazine,* 8(2), 36–40, 2002.
5. U.S. Department of Agriculture, Belgium/Luxembourg Sanitary/Phytosanitary/Food Safety Quality and Traceability Concerns Spread to Vegetable Producers Chain, Foreign Agricultural Services GAIN Report #BE1025, June 29, 2001.
6. United Fresh Fruit and Vegetable Association, Food Safety Questionnaire for Fresh Fruits and Vegetables, 3rd draft, 2001.
7. Stier, R.F., GMPs and HACCP for Beverages, short course sponsored by the Institute of Food Technologists, 1998.

2 The Role of Genetically Modified Organisms (GMOs) in Beverage Production

Susan Harlander

CONTENTS

History of Genetic Modification of Food Plants and Animals....................10
Regulation of Genetically Modified Crops 10
Identity Preservation and the International Market 11
    Detection of Genetically Modified Ingredients................................12
    Difficulties with Product Labeling 12
The Future of Genetically Modified Foods 13

In the relatively short time since their commercial introduction in 1996, genetically modified (GM) crops have been rapidly adopted in the U.S. The first products of plant biotechnology involve input traits, such as herbicide tolerance and insect resistance. Of the 51 products reviewed by the U.S. Food and Drug Administration (FDA), the vast majority are commodity crops such as corn, soybeans, and canola. Because FDA considers these crops "substantially equivalent" to their traditional counterparts, no special labeling is required for GM crops in the U.S., and they are managed as commodities with no segregation or identity preservation (IP). This creates an issue for multinational beverage manufacturers since labeling guidelines for and consumer acceptance of GM crops differ in other parts of the world. This chapter will focus on the challenges associated with establishing IP systems for commodity ingredients through a food supply chain geared for maximum efficiency and least cost. It will also address current testing systems for GM ingredients, including both protein- and DNA-based methods. The growing need for accurate, specific, reliable, standardized, and validated testing methods to ensure compliance with established threshold levels for GM ingredients as well as global labeling guidelines will be discussed. Finally, examples of next-generation biotechnology products of relevance to the beverage industry will be provided.

HISTORY OF GENETIC MODIFICATION OF FOOD PLANTS AND ANIMALS

People have been genetically modifying the food supply during the thousands of years since the domestication of plants and animals began. Classical breeding and selection, as well as techniques such as radiation breeding, embryo rescue, and transposon mutagenesis, create significant changes in the genetic makeup of plants and animals due to the random recombination and sorting of thousands of genes. As a result of intervention by people, the hybrid seed corn currently grown throughout the world bears little resemblance to teosinte, the original ancestor of corn. The newer techniques involving genetic engineering, on the other hand, allow for the transfer of a few genes in a much more precise, controllable, and predictable manner than that occurring as a result of conventional breeding. Interestingly, plants improved through conventional genetic modification methods undergo no formal food or environmental safety evaluation prior to introduction into the marketplace, whereas genetically engineered crops are required to undergo extensive food and environmental safety testing before their introduction.

Genetically modified crops were first commercially introduced in the U.S. in 1996 and have been rapidly adopted by farmers. It has been estimated that 24% of the corn and almost 70% of the soybeans and cotton grown in the U.S. in 2001 were GM varieties. Examples of GM crops include insect-resistant (Bt) corn, cotton, potato, and tomato; herbicide-tolerant soybeans, corn, rice, sugar beet, flax, and canola; and virus-resistant squash, papaya, and potato. Advantages of insect- and virus-resistant crops include improved yields and reduced use of pesticides. An additional benefit of Bt corn is reduced contamination by fumonisin-producing fungi. Fumonisin is a potent mycotoxin implicated in esophageal cancer and neural tube birth defects in humans. Advantages of herbicide-tolerant crops include improved weed control, reduced crop injury, reduction in foreign matter, reduced fuel use, and significant reduction in soil erosion. It is for these reasons that GM crops are the most rapidly adopted technology in the history of agriculture.

REGULATION OF GENETICALLY MODIFIED CROPS

GM crops are regulated in the United States through a coordinated framework developed in 1992 and administered by three agencies: the U.S.

Department of Agriculture (USDA), the Environmental Protection Agency (EPA), and the FDA. Rigorous food and environmental safety assessments must be completed before GM crops can be commercialized. An effective food safety evaluation system minimizes risk, but it is important to remember that food is not inherently safe. There are numerous examples of natural toxicants present in various foods (e.g., solanine in potatoes and glycoalkaloids in broccoli). If we were to eliminate all foods that posed any kind of risk, our food choices would be very limited. The goal of a food safety system is "reasonable certainty of no harm" at normal levels of consumption. Acceptance of a new food product occurs when it is shown to be as safe as or safer than its conventional counterpart; therefore, the final assessment of safety is always comparative.

The scientific basis of the evaluation process is the concept of "substantial equivalence." Regulatory agencies compare GM crops to their conventional counterparts. A wide range of comparisons is made including nutritional equivalency, levels of natural toxicants, and the potential for allergenicity, in addition to a number of agronomic and environmental factors. If the GM crop is essentially identical to its conventional counterpart in all aspects, it is considered substantially equivalent, and no special labeling is required in the U.S. Over 400 million acres of GM crops have been grown worldwide, and there has not been a single documented adverse health effect or food safety issue associated with consumption of these products.

Since GM crops are substantially equivalent and no labeling is required, they have been managed as commodities in the U.S. and have made their way through commodity distribution channels into thousands of ingredients used in processed foods. It has been estimated that greater than 70% of all processed foods contain one or more ingredients potentially derived from GM soy or corn. Examples of soy- and corn-derived ingredients found in beverages include cornstarch, corn syrup, corn syrup solids, dextrose, high-fructose corn syrup, soybean oil, and lecithin. Genetic engineering has also been used to produce vitamins and flavors, and many milk-derived ingredients used in beverages have been derived from cows treated with recombinant bovine somatotropin.

IDENTITY PRESERVATION AND THE INTERNATIONAL MARKET

In the past, it was not necessary for the food supply chain to segregate and identity preserve grain destined for ingredient manufacture. However, several countries have adopted labeling guidelines for foods containing ingredients derived from GM crops. Because GM foods are perceived negatively in these countries, food manufacturers try to avoid GM ingredients in order to avoid labeling their products. Unfortunately, the infrastructure of agriculture has not yet evolved to the stage where it can deliver large quantities of IP grains. When available, IP grains are more expensive than their conventional counterparts due to the added labor and costs associated with segregation, quality control, and testing. Comingling of GM with non-GM crops at any stage in the food ingredient chain from seed to final product could potentially result in mislabeled products and significant liability for the food and beverage industries.

DETECTION OF GENETICALLY MODIFIED INGREDIENTS

To authenticate label claims, food processors need standardized and validated analytical methods for detecting the presence of GM ingredients. Unfortunately, standardized methods do not currently exist for most of the GM ingredients on the market today. Two types of tests are used for the detection of GM material. The first method involves enzyme-linked immunosorbent assays (ELISAs), which are based on the detection of proteins coded for by the genes inserted into GM crops. These tests require minimal sample preparation and are sensitive, accurate, rapid, and inexpensive. They can only be used on unprocessed samples, however, as proteins are denatured by heat and other food processing methods. The second method is based on direct detection of the gene(s) (DNA) inserted into GM crops. The DNA is typically amplified using polymerase chain reaction (PCR) technology to increase the amount of DNA to detectable levels. PCR methods require extensive sample preparation, the procedure is lengthy, and per sample costs are high. The method is very sensitive and can be used to detect DNA in processed samples.

The current methods for detecting GM material in foods have numerous limitations. Authenticated reference standards are not available, and every laboratory has developed its own testing protocols. False positive and false negative rates are unacceptably high. No standardization of how the results are reported to food and beverage companies has been developed. The food matrix has a dramatic impact on extractability of DNA and protein, and protocols will need to be developed to take this into account. Since labeling is not required in the U.S., detection methods have not developed as rapidly as GM technology. This deficiency will cause significant issues as disputes about the GM status of foods arise. Several efforts are currently underway to validate and standardize GM testing methods, but to date, only one ELISA for herbicide-tolerant soybeans has been validated and standardized.

DIFFICULTIES WITH PRODUCT LABELING

Despite these challenges, some companies are overtly labeling their products as GMO-free or non-GM. They procure ingredients from suppliers who certify that non-GM varieties have been used for ingredient manufacture. A recent report in the *Wall Street Journal* (April 2001) stated that of 20 products labeled as non-GM, 16 contained measurable quantities of GM DNA. Therefore, even under best-case scenarios, it is very difficult to guarantee that the non-GM label is truthful.

Most U.S. food companies are not avoiding GM ingredients for domestic production. In general, the U.S. food processing industry has confidence in the safety of GM foods. Because GM crops have been readily adopted in the U.S., availability of non-GM crops has been limited, and these ingredients are more expensive. Even when efforts are made to procure non-GM ingredients, adventitious contamination is an issue, and IP systems have not been perfected, as was illustrated with the StarLink™ incident in 2001. The food industry would need to be able to accurately forecast its supply needs for non-GM ingredients so farmers could be instructed on the quantities required. In addition, the food industry lacks the separate storage, processing, labeling, and transportation capabilities required to ensure separation of GM and non-GM raw materials and final products. Little confidence exists in the adequacy of current GM sampling and testing methodology to substantiate label claims, and substantial liability exists if label claims are inaccurate. Consumers of processed foods in the U.S. do not appear to be overly concerned about the presence of GM ingredients. Food manufacturers have been monitoring their 800 numbers for an indication of how their consumers feel about GM foods. To date, the number of calls on biotechnology remains very small (0.1 to 0.2%) for most major food companies in the U.S.; however, awareness remains relatively low. Calls increase during periods of intense media coverage, and companies targeted by activist groups report periodic increases in numbers of calls. If a brief explanation of biotechnology is provided, acceptance increases significantly, indicating that education is an important factor in consumer acceptance. Finally, the food and beverage industries hope that the next generation of GM products will deliver compelling consumer benefits.

THE FUTURE OF GENETICALLY MODIFIED FOODS

The next generation of GM foods will focus on "output traits" that provide tangible consumer-relevant benefits. Biotechnology can be used to remove allergens, natural toxicants, and antinutrients from foods such as peanuts, soybeans, rice, and wheat. Taste, texture, aroma, ripening time, and shelf life of fresh fruits and vegetables can be improved. It will be possible to improve the nutritional quality of foods. Examples include modification of the saturation level of oils to produce products high in monounsaturated fatty acids that are more stable, resist oxidation, do not require hydrogenation, and reduce cholesterol levels when consumed in place of saturated fatty acids. It is possible to increase the content of vitamin E, a natural antioxidant, and to insert the capability of producing plant-based omega-3 fatty acids into oil seeds. Biotechnology can be used to elevate levels of vitamins A, C, and D and folate; increase antioxidants; and enhance iron bioavailability in vegetables, fruits, and grains. It is also possible to increase the levels in various plants of phytochemicals that have been associated with disease prevention, e.g., lycopene in tomatoes and sulfurofane in broccoli for reducing cancer risk, lutein in vegetables for reducing risk of macular degeneration, etc. The advancing fields of human and plant genomics and proteomics will identify additional plant-based compounds that could have a positive impact on human health. These are the kinds of products that will excite food and beverage companies and ultimately consumers in the future.

3 Beverages as Delivery Systems for Nutraceuticals

Dennis T. Gordon and Kiyoko Kubomura

CONTENTS

Introduction 15
Defining Nutraceuticals/Functional Foods 18
Beverages — Liquid Foods 19
Classes of Nutraceuticals...45
Biochemical, Physiological, and Molecular Actions of Nutraceuticals 53
Conclusion and Future Considerations 59

INTRODUCTION

The concept of nutraceuticals or functional foods is nothing short of an awakening. Heasman and Mellentin aptly titled their book *The Functional Foods Revolution, Healthy People, Healthy Profits?*[1] The authors discuss the origin and development of the concept and provide a fascinating account of food product development and marketing techniques for health promotion. They also write a monthly publication, *New Nutrition Business,* which chronicles advances and setbacks in this dynamic field of foods for health (see www.new-nutrition.com). As a relatively new idea, the marketing of nutraceuticals or functional foods is far outpacing available science in an attempt to prove efficacy. However, we are convinced that the nutraceutical/functional food revolution is real and important. In the long term, this concept is likely to expand food science, play a major role in the nutrition of the twenty-first century, and represent new horizons for human development and health. As with all new science, the spin-off success stories may outdistance the original idea.

The terms nutraceuticals and functional foods are synonymous. However, many experts in this field prefer nutraceuticals, for reasons first proposed by

Stephen DeFelice.[2] The term nutraceuticals is used in the title of this chapter, and an explanation for the preference for the term is found in this question: does a single food (functional food) contribute to health and disease prevention or is it one or all of the chemical compounds working in conjunction in foods (functional food ingredients or nutraceuticals) that contribute to health and disease prevention? With this question in mind, the term nutraceuticals relates better to the chemical compounds that have the biochemical, physiological, and molecular functions that contribute to health. Conversely, it is specific foods or combinations of foods that have shown positive correlations with the reduced incidence of diseases in epidemiological studies.[3-5] Recommendations for consumption or avoidance of specific foods and changes in dietary patterns receive a great deal of support based on epidemiology. Ultimately, clinically based experimental studies are needed to prove the efficacy of nutraceuticals.[6-8] We are reminded that all foods are functional foods and contain a variety of nutraceuticals, although at times we isolate or concentrate individual nutraceuticals as direct supplements or as additions to solid foods or beverages.[9] The total importance or lack of significance of the many nutraceuticals is not known. Nor do we know the importance of the interactions among nutraceuticals and other food components. The science of nutraceuticals is a dynamic, new discipline. The term nutraceutical will be used throughout this review.

Many beverage products have had tremendous consumer acceptance as attempts have been made to associate consumption with improved health, performance, stamina, mood, or general state of well being. Although these products have used catchy marketing names and mixtures of vitamins, minerals, botanicals, herbs, or other supplements, most of them lacked adequate scientific data to support their claims. In many instances, claims were made that the beverages provided instant relief or satisfaction, but clear knowledge about the purity and efficacy of the ingredients used in the beverages was lacking. With regard to some nutraceutical beverages on the market today, the best advice for the consumer is still "to be aware." This review is intended to help foster the development of nutraceutical beverages based on science rather than testimonials, marketing slogans, and product names alone.

Today, the consumer is more interested in health than nutrition.[10] Consumers' willingness to purchase foods that might provide for improved health has created a marketing bonanza for the food industry and an awakening for the scientific community. Yes, essential nutrients can improve health and prevent disease, but the number of star essential nutrients for successful marketing and improved food sales is, at present, limited. Calcium builds strong bones, but it can also prevent osteoporosis — one of the top 10 chronic diseases in the United States.[11] Folic acid is essential for the transfer of one-carbon (methyl) units in the biosynthesis and metabolism of amino acids, nucleotides, and other cellular molecules.[12] However, the acceptance of folic acid is more easily grasped by the consumer as an aid in the prevention of neural tube birth defects.[13] The marketing of foods containing added essential nutrients, such as calcium or folic acid, has been a true success story for both public health and the food industry. Today, however, the topic of nutraceuticals is much broader than the essential nutrients that relate foods to health for the consumer.

The original discipline of food science stressed improvement of the attributes of food quality, including taste, texture, aroma, safety, color, and nutrition, by ensuring nutrient stability and content. Research by nutritionists emphasizes essential nutrient discovery, function, and the establishment of appropriate recommended dietary allowances (RDAs),[14] now known as Dietary Reference Intakes (DRIs).[15] Today, both professional disciplines focus on all the ingredients in foods (nutraceuticals) for disease prevention and longer, more productive lives. The food scientist has the added challenge of understanding the impact of nutraceuticals on food quality and taste.[16] It is estimated that approximately one-third of all current funding for research and product development in the food industry is applied to health-related foods.[17]

Essential nutrients are necessary for growth and maintenance of bodily functions. There are 41 essential nutrients: water, 11 amino acids, two fatty acids, 14 vitamins, and 13 minerals. There is continued discussion about whether some ultra-trace elements should be defined as essential (i.e., B, Cd, Ni, and V). However, foods contain over 10,000 other compounds, called nutraceuticals, which are being extensively investigated for their possible health effects. Not all these compounds can be covered in this review, and it remains to be determined how many will prove to be important in human metabolism and health. Simply put, the importance of nutraceuticals and the mechanisms by which nutraceuticals could complement essential nutrients for growth and maintenance is not known. This review does not cover essential nutrients that are frequently promoted as functional food ingredients (e.g., calcium and folic acid). However, this review can be viewed as a broad primer that relates nutraceuticals in foods, specifically liquid foods and beverages, to improved health. It remains a challenge to determine all those liquid foods that can fall under the umbrella term of beverages. This review is also intended to help the reader categorize the nutraceuticals found in foods into nine classes and cites examples of individual nutraceuticals along with their proposed beneficial function and efficacy in the body. Many nutraceuticals have multiple functions in health promotion.

The following main topics are discussed with the objective of integrating nutraceuticals and beverages for this review: defining nutraceuticals/functional foods; beverages — liquid foods; classes of nutraceuticals; biochemical, physiological, and molecular actions of nutraceuticals; and conclusion and future considerations.

DEFINING NUTRACEUTICALS/FUNCTIONAL FOODS

There are no official U.S. or international definitions for functional foods or nutraceuticals. A useful working definition proposed by the U.S. Institute of Medicine is "any modified food or food ingredient that may provide a health benefit beyond the traditional nutrients it contains."[18] Nutraceuticals have been defined as "naturally derived bioactive compounds that are found in foods, dietary supplements, and herbal products, and have health promoting, disease preventing, or medicinal properties."[19] There is an ongoing discussion internationally about how the concept of functional foods should be described and regulated. A formal definition would imply an acceptance of the principle, and recognition to some degree, of one or more functional foods. The U.S. Food and Drug Administration (FDA) does not have a definition for nutraceuticals but regulates these foods under the authority of the Federal Food, Drug, and Cosmetic Act.[20] In so far as a nutraceutical is a dietary supplement, it is regulated by the FDA under the Dietary Supplement Health and Education Act (DSHEA).[21] This law allows the use of structure/function claims. However, these claims cannot be related to a disease. An example of an FDA-impermissible structure/function claim for chondroitin sulfate (for joint inflammatory disorders) would be "reduces the pain and stiffness associated with arthritis," but the FDA would probably allow the claim "helps build and strengthen joint cartilage."

Japan, specifically the Otsuka Pharmaceutical Company, is recognized for initiating the concept of functional foods with the introduction of their product Fibre Mini, a beverage. Dietary fiber, specifically soluble dietary fiber, became a worldwide nutritional phenomenon in the mid-1980s.[22] Polydextrose, a low-molecular-weight nondigestible carbohydrate, was the ideal ingredient as a source of soluble dietary fiber.[23] Five grams of polydextrose were added to 100 ml of water with coloring and flavorings. In Japan, where the population is keenly interested in the relationships between foods and health, Fibre Mini was and remains a success.

In 1991, Japan moved away from the term functional foods and introduced the concept of FOSHU (Foods for Specified Heath Use). FOSHU represents a collaboration between the food industries of Japan and the Japanese government for self-regulation of food products that promote specific health messages. Again, it is the ingredient, the nutraceutical added to a "food for a specified health use," that is being promoted. Based on information supplied by the Matsutani Chemical Company (H. Okuma, personal communication), as of May 2002, there were 295 foods approved as FOSHU. These foods, or more specifically the nutraceuticals they contain, are divided among 10 different physiological conditions or specific health uses and are listed in Table 3.1. Within each category for "specific health use," there are approved individual compounds or fractions of foods or bacteria allowed and promoted as nutraceuticals. These include 46 chemicals or extracts and 16 microorganisms (probiotics). Many of the approved uses for these nutraceuticals are in beverages. The information presented in Table 3.1 can be viewed as an introduction to a variety of nutraceuticals and their uses to improve human health. Examples of FOSHU-approved nutraceuticals (Table 3.1) are cited in this review. However, the nutraceuticals cited in Table 3.1 and the science supporting their efficacy in health promotion deserve greater evaluation than can be afforded here. Table 3.2 divides nutraceuticals into nine classes based on their manufacture or simple chemical composition and characteristics in foods. Table 3.3 attempts to list some of the major classes of biochemical, physiological, and molecular actions of nutraceuticals in the body. The information in Table 3.1, Table 3.2, and Table 3.3 is complementary.

Numerous books and proceedings on nutraceuticals and functional foods are available. Only a few are cited.[1,24-27]

BEVERAGES — LIQUID FOODS

While all foods nourish, most foods, with the major exceptions of water, milk, and alcoholic beverages, initially exist as solids. Theoretically, all solid foods can be delivered in a liquid form. Soups are the ultimate example of using any food or combination of foods to make a meal; soups can be served as cold or hot beverages. Homemade chicken soup has always been perceived to be healthful.[28] Before the concept of nutraceuticals was introduced, chicken soup competed with wine and yogurt for perceived health-giving properties. The nutraceuticals, if any, in chicken soup are unknown, but wine is rich in phenolics. Yogurt contains starter culture organisms, and some probiotics are also added to yogurt.[29] The only limitations to a food in a liquid form, such as a soup or beverage, are technology and consumer acceptance. Beverages are accepted by the consumer, are convenient, and can be marketed to meet consumer demands for container contents, size, shape, and appearance. Faced with the conundrum of how best to describe beverages, Figure 3.1 attempts to list the categories of liquid foods and beverages that can or have been described as nutraceutical beverages. When some types of beverages fall into multiple classes, as listed in Figure 3.1, the question can be asked, what is not a beverage?

Use of the term "juice" has legal ramifications. To legally be called a juice, the product's liquid must contain no less than 100% of that food.[30]

TABLE 3.1
Nutraceuticals (Functional Food Ingredients) Used in Foods Approved for Specific Health Use (FOSHU) in Japan[a]

INGREDIENTS[b]	APPLICATIONS[c]
I. Intestinal regularity (171)	
A. Dietary fiber	
1. Indigestible dextrin (Fibersol-2) (20)	Soft drinks
2. Psyllium (19)	Powdered drinks
3. Hydrolyzed guar gum (4)	Drink-type yogurt
4. Polydextrose (2)	Soup
5. Wheat bran (4)	Potage
6. Depolymerized sodium alginate (2)	Rice porridge
7. Dietary fiber from beer yeast (1)	Cereals
8. Dietary fiber from agar (3)	Sausage
	Precooked rice noodles
B. Oligosaccharides	
1. Lacto-fructo-oligosaccharide (24)	Soft drink
2. Fructo-oligosaccharide (11)	Powdered drink
3. Soy-oligosaccharide (7)	Table sugar
4. Xylo-oligosaccharide (5)	Tablet candy
5. Galacto-oligosaccharide (7)	Candy
6. Isomalto-oligosaccharide (3)	Cookie
7. Lactulose (1)	Chocolate
8. Lavinose (1)	Pudding
	Syrupy aloe
	Frozen yogurt
	Vinegar
	Tofu
C. Dietary fiber and oligosaccharide	
1. Galacto-oligosaccharide and polydextrose (1)	Soft drink
D. Lactic acid bacteria	
1. *Lactobacillus casei* Shirota (26)	Drink-type yogurt
2. *Bifidobacterium bleve* Yakult (4)	Yogurt
3. *Lactobacillus delbrueckii* subsp. *bulgaricus* 2038 and *Streptococcus salivarius* subsp. *thermophilus* 1131 (6)	Lactic acid bacteria drink
4. *Bifidobacterium longum* BB536 (5)	
5. *Lactobacillus* GG (2)	
6. *Lactobacillus acidophilus* ABT-2062 and *Bifidobacterium longum* SBT-2928 (1)	
7. *Bifidobacterium lactis* FK 120 (2)	
8. *Bifidobacterium lactis* LKM512 (2)	

TABLE 3.1 (CONTINUED)
Nutraceuticals (Functional Food Ingredients) Used in Foods Approved for Specific Health Use (FOSHU) in Japan[a]

INGREDIENTS[b]	APPLICATIONS[c]
9. *Bifidobacterium acidophilus* CK 92 and *Lactobacillus helveticus* CK60 (5)	
10. *Lactobacillus casei* NY 1302 (1)	
11. *Lactobacillus gaseri* sp. and *Bifidobacterium bifidus* sp. (1)	
12. Propionic acid bacterium (1)	
II. For people with high cholesterol levels (28)	
A. Mixed chemical and physical properties and sources	
1. Soy protein (15)	Soft drink
2. Depolymerized sodium alginate (4)	Powdered drink
3. Chitosan (4)	Soy milk
4. CSPHP – Soybean-protein-hydrolysate with phospholipids (2)	Cookie
	Yogurt
5. Plant sterol esters (1)	Fried bean curd cake
6. Plant stanol esters (1)	Sausage
7. Plant sterols (1)	Hamburger
	Meatball
	Precooked Chinese noodles
	Margarine
III. Intestinal regularity and for people with high cholesterol levels (9)	
A. Dietary fiber	
1. Depolymerized sodium alginate (6)	Powdered drink
2. Psyllium (3)	Soft drink
IV. For people with high blood pressure (23)	
A. Nitrogen compounds	
1. Sardine peptide (Valyl-tyrosine peptide) (10)	Soft drink
2. Lacto-tri-peptide (2)	Powdered soup
3. *Katsuobushi* (bonito) oligo-peptide (6)	Supplement
4. Casein dodecanoic peptide (3)	
B. Phenolic	
1. *Tochucha* herb tea glycoside (2)	
V. Promotes mineral (calcium, iron) absorption (21)	
A. Mixed chemical and physical properties and sources	
1. Calcium phosphopeptide (CPP) (3)	Soft drink
2. Calcium citrate malate (CCM) (2)	Soy milk

(continued)

TABLE 3.1 (CONTINUED)
Nutraceuticals (Functional Food Ingredients) Used in Foods Approved for Specific Health Use (FOSHU) in Japan[a]

INGREDIENTS[b]	APPLICATIONS[c]
3. Heme Fe (4)	Tofu
4. Fructo-oligosaccharide (FOS) (5)	Natto
5. *Bacillus subtilis* OUV23481 (Vitamin K_2) (3)	
6. Soy isoflavones (3)	
7. Milk basic protein (MBP) (1)	
VI. Prevent dental caries (13)	
A. Mixed chemical and physical properties and sources	
1. Maltitol (2)	Chewing gum
2. Palantinose and tea polyphenol (1)	Chocolate
3. Maltitol and palantinose and tea polyphenol (1)	Candy
4. Maltitol and palantinose and erythritol and tea polyphenol (1)	Tablet
5. Casein phosphopeptide–amorphous calcium phosphate compound (CPP-ACP) (6)	
6. Xylitol and hydrogenated palatinose and calcium phosphate and *Gloiopeltis furcata* (seaweed) extract (2)	
VII. For people who care about high blood glucose levels (20)	
A. mixed chemical and physical properties and sources	
1. Indigestible dextrin (Fibersol-2) (14)	Soft drink
2. Wheat albumin (3)	Powdered drink
3. Guava leaves polyphenol (1)	Powdered soup
4. L-Arabinose (1)	Freeze-dried miso soup
5. *Touchi* (fermented black beans) extract (1)	Tofu
VIII. Inhibits postprandial increase of serum triglyceride levels and prevents fat accumulation and for people with high cholesterol levels (4)	
A. Mixed chemical and physical properties and sources	
1. D-Acyl-glycerol and plant sterol (b-cytosterol) (4)	Cooking oil
IX. Inhibits postprandial increase of serum triglyceride levels and prevents fat accumulation	
A. Lipid	
1. D-Acyl-glycerol	Cooking oil

TABLE 3.1 (CONTINUED)
Nutraceuticals (Functional Food Ingredients) Used in Foods Approved for Specific Health Use (FOSHU) in Japan[a]

INGREDIENTS[b]	APPLICATIONS[c]
X. Inhibits postprandial increase of serum triglyceride levels (3)	
A. Nitrogen compound	
1. Globin hydrolysate (3)	Soft drink
	Jelly-type drink

[a] Nutraceuticals are listed among 10 groups for specific health use. Some nutraceuticals have multiple approved specific health uses.
[b] Number in parentheses indicates number of products approved with this ingredient.
[c] Examples of products using these ingredients.

Beverage is a generic term, which can include many liquid foods. A beverage that purports to contain fruit or vegetable juice must bear on the information panel of the label a statement of the total percentage of juice.[31] The naming of a nonstandardized juice beverage must comply with common or usual name regulations.[32] If water is added to dilute the juice, the name must include a term such as "drink," "beverage," or "cocktail." It is not the intent of the authors of this review to invent nomenclature for liquid foods or beverages, and the authors accept all responsibility if any formal or legal rules have been violated in the proposed classification system (Figure 3.1).

Beverages can acquire the same accolades associated with solid foods, such as nourishment, enjoyment, relaxation, performance, and health. Beverages can also be described as nutraceuticals. However, nutraceutical foods or beverages are not drugs. And there is a tendency to think of nutraceuticals

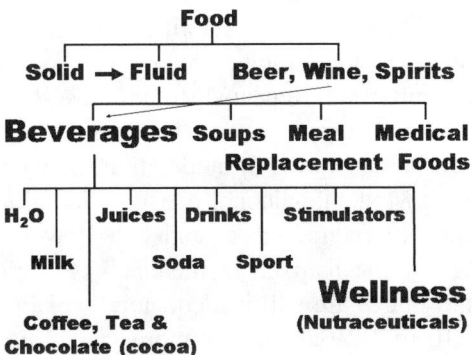

FIGURE 3.1 Categories of liquid foods and beverages.

as having drug-like properties that produce the accelerated health improvements obtained with many prescribed drugs. This association of health foods or foods for health with drugs helped foster the term "pharmafood."[33] Accumulating evidence suggests that nutraceuticals contribute to health, but it is through their consumption, in a varied diet, over a long period of time — possibly a lifetime. It is also important to remember that overconsumption of any food or nutraceutical, in particular certain botanicals, can be harmful, and in some cases more harmful than abstinence.[34] Because many of the food ingredients promoted for nutraceutical properties are relatively new to the diet, information on the Tolerable Upper Intake Level (UL) of any nutraceutical is totally lacking. The National Academy of Sciences defined the UL of a nutrient to be the highest level of daily intake that is likely to pose no risk of adverse health effects for a high percentage of the population. As the intake of a nutrient increases above the UL, the potential risk of adverse affects increases.[35] The UL for some essential nutrients remains to be established.[35] The botanical kava-kava is an example of a nutraceutical for which safety concerns exist. Kava-kava received a high degree of consumer acceptance as a relaxant. Now, sufficient evidence has shown that kava-kava is associated with liver toxicity; it is described as unfit for human consumption in the U.K.[36] Prior to preliminary warnings about kava-kava and a formal statement about its toxicity,[36] it was added to beverages and marketed without safety tests or UL investigation.

Beer, Wine, and *Spirits* are products of cereals, fruits, and potentially any plant food that could be fermented with yeast to yield alcohol. Ciders could be added to this group. It is arguable which is the most important or beneficial to the consumer, the unique taste of the beer, wine, or spirit or its alcohol content. Both are important in the context of nutraceuticals. Except for spirits, most fermented plant-based beverages contain significant amounts of phenolics, a major class of functional food ingredients. Phenolics make significant contributions to the taste of beer and especially to the taste of wine.[37] Alcohol is energy dense, containing 7 calories (kcal) per gram, compared to carbohydrate and protein, which have 4 kcal per gram, but less than the 9 kcal in a gram of fat. Repeated studies have shown the therapeutic value of moderate alcohol consumption.[38] The U.S. Dietary Guidelines suggest that alcohol can be consumed in moderation; they do not say to avoid alcoholic beverages.[39] Moderate alcohol consumption along with a prudent diet has been shown to reduce stress and help lower blood cholesterol levels.[40,41] However, the mechanisms of the changes in blood lipid concentrations with alcohol intake are still inadequately explained.

Wine contains 10 to 12% alcohol or more, but it also contains a high concentration of phenolic compounds. The process of winemaking concentrates the phenolics derived from grapes, especially in the production of red wine. Red wines can contain 1000 to 3000 mg/l of phenolic compounds, compared to white wines, which contain approximately 200 mg/l.[42] All the major phenolics in wine appear to have antioxidant properties, but to varying degrees. Initial attention focused on resveratrol in red wine as possibly contributing to lower coronary heart disease (CHD) mortality. This suggestion was based on a study conducted in hyperlipidemic rats in which resveratrol was shown to reduce platelet aggregation and lower blood cholesterol levels.[43] Resveratrol has also been shown to be an antioxidant capable of protecting the lipids in low-density lipoproteins (LDL) in blood against oxidation. It has been suggested that this particular capability is the mechanism to explain the reduced incidence of CHD among the French population; thus, the French paradox.[44] However, other more abundant flavonoids (i.e., epicatechin and quercetin) have since been identified in wine and are suspected of being more significant than resveratrol in serving as (nutraceuticals) antioxidants in the body.[45] Beers also contain phenolics, but in smaller amounts and of different chemical composition than the phenolics in red wines.[46] Ciders and apple juice also contain phenolics, but they are different from those in wine and beer.[47]

Strong connections exist between the foods of a region and health. These associations are continually being investigated. The association between wine consumption and the reduced risk of CHD helped coin the term "French paradox." Similarly, the high consumption of tea and soy among the Japanese and their comparatively lesser incidence of various diseases introduced the term "Japanese paradox." The term "Mediterranean diet" is based on the comparatively lower rates of heart disease and other diseases among individuals living in regions where the diet is high in olive oil, fruits, and vegetables.

Fluid Meal Replacements offer convenience and can almost be considered capable of meeting complete nutritional needs for short to moderately extended periods of time. Products in this category include Ensure® and Slimfast®. Although these products are not considered main-line nutraceutical foods or beverages, they contain soy protein, a highly regarded nutraceutical protein known to lower blood cholesterol levels,[48] and the phenolics genistein and daidzein (genistin and daidzin are the glycoside forms naturally occurring in soy), thought to help prevent breast cancer and other disorders.[49] The current food label claim for Ensure states that the product contains "complete, balanced nutrition to help stay healthy, active and energetic." The front panel also contains the words, "Now! Lutein to help support eye health." Lutein is a xanthophyll (tetraterpenoid) that acts as an antioxidant and is believed to be associated with the prevention of cataracts and age-related macular degeneration (ARMD).[50,51] According to its label claim, each serving of Ensure (8 fl oz.) has 500 µg of lutein. A trend among food companies is to list the amount of a nutraceutical provided by a serving. Fluid Meal Replacements, for convenience or weight control, are excellent examples of nutraceutical beverages that can be modified to contain many different nutraceuticals.

Medical foods or enteral formulas are provided in liquid form and are usually prescribed by a physician for specific diseases or disorders. Medical foods are primarily intended for patients in hospitals or for individuals with rare diseases and can include a broad range of products and ingredients. Some of the specialized nutrients or nutraceuticals used in medical foods include protein and amino acids, branched-chain amino acids, glutamine, carnitine, taurine, ribonucleic acid (RNA), fatty acids and medium chain triglycerides, and dietary fiber. The topic is well reviewed in the Institute of Food Technologists (IFT) Scientific Status Summary entitled Medical Foods.[52] The fact that medical foods are prescribed and evaluated with medical supervision will help provide the science needed to show the efficacy of some nutraceuticals. Thus, some of these nutraceuticals will eventually find their way into more mainstream functional foods for the health-conscious consumer.

It is possible that *Water, H_2O,* could be listed first among the 41 essential nutrients. All life processes evolve through this aqueous environment. The human body is 60% water,[53] and most foods except cereals and grains contain high levels of water. In the 1990s, dietitians and other health professionals encouraged a greater intake of water, but today, occasional cautionary notes to avoid water toxicity are seen.[54] Still, in the last several years, the sale and consumption of bottled water products has been one of the most important and significant phenomena to affect consumer nutrition, health, and profits for the food industry.[55] Is this a social phenomenon or a move by consumers to drink more water (fluid) as they become more health conscious? Followed by milk, water is possibly the most natural and quintessential nutraceutical beverage. The increase in bottled water sales might also reflect an increased emphasis on exercise. Exercise is vital to good health, and even moderate exercise requires a person to be properly hydrated.

Water is usually the first vehicle of choice for delivery of any nutraceutical or supplement to make a beverage, drink, or cocktail. Milk and juices also receive their share of added nutrients and nutraceuticals. However, for almost all the nutraceuticals, either as they exist naturally in foods or as additives to water, relatively little is known about efficiency of absorption into the bloodstream, assimilation into organs or tissues, or efficacy for a specific disease or disorder. Yes, there are exceptions to this general statement,[56–59] and research on the bioavailability of nutraceuticals is rapidly increasing. This cautionary statement refers to the earlier statement that much remains to be accomplished in the science of nutraceuticals. The combination of energy compounds, electrolytes, stimulators, and nutraceutical agents that can be added to water is unlimited. Equally unlimited are the names used to market and promote these products (i.e., Fortified Water, Power Water, Vitamin Water, Fitness Water). These descriptively named products lack scientific support in most cases. In fact, the story of nutraceutical beverages has not always been one based on solid science because of the indiscriminate addition of a host of nutraceuticals to water.

Milk can certainly be described as the single most important food and the only food in the diet of the newborn. Cow's milk should and will continue to receive intense investigation as a functional food.[60] Lactoferrin in cow's milk has received attention as an intestinal antimicrobial agent through its ability to chelate iron, which prevents it from being available to allow pathogenic bacteria to multiply in the newborn's intestine.[61] Endogenous galactooligosaccharides (GOS) in human breast milk have been shown to bind to pathogens in the newborn's intestine, which prevents their adherence to their intestine,[62] and it is known that the GOS-pathogen complex is passed to the large intestine. After GOS reaches the large intestine, it is fermented and serves as a prebiotic,[63] which helps maintain a more acid environment and a healthier bacterial population in the infant's colon. Both lactoferrin and GOS are excellent examples of nutraceuticals, naturally occurring in cow's milk and in human milk, respectively. Lactoferrin could be used in nutraceutical beverages, and GOS is used in nutraceutical beverages. A manufactured source of GOS is available and is added to beverages as a prebiotic in Japan (Table 3.1). Milk is frequently supplemented with vitamins, minerals, and now nutraceuticals, too (i.e., probiotics). Although milk was often described as possibly nature's most perfect food, its image was unfairly tarnished because of its natural abundance of cholesterol and saturated fat and the association of these compounds with CHD. The issue has been addressed by the combined food and dairy industries by making lowfat or nonfat dairy products available; these are now popular beverages in many households. Other milk-derived beverages that are receiving attention as nutraceuticals include kefir,[64] colostrums,[65] and yogurts.[66] These fermented foods and other liquid dairy products are the vehicles of choice when probiotics are added.[67] Although chocolate milk may have been regarded as a sweet liquid snack, current knowledge about the therapeutic benefits of chocolate with its varied phenolic contents would elevate chocolate milk to a nutraceutical beverage. The health benefits of phenolics in cocoa and chocolate will be discussed later in this chapter. The nutraceutical potential of cow's milk is an exciting story, with much that is still to be discovered.

One important dietary change due to the interest in nutraceuticals is the introduction of soy milk[68] and other nonbovine milk-related beverages, such as rice milk. Again, soy serves as a functional food because of its protein, which has been shown to lower blood cholesterol,[69] and its endogenous phenolic compounds, which have antiestrogenic properties reported to reduce the incidence of breast cancer[70] and the symptoms of postmenopausal syndrome.[71] Rice-based beverages are beneficial for those individuals (infants) who have allergies to bovine milk and or soy milk.[72]

Coffee, Tea, and *Cocoa (Chocolate)* are beverages that provide enjoyment and minimal nourishment but are stimulants because of their caffeine content. Caffeine is considered more as a flavor enhancer, stimulant, or energizer than as a nutraceutical directly related to a disease or to better health. Although it is a controversial issue, possibly because of its popularity, the consumption of coffee has not been found to be harmful[73-75] and may be beneficial. However, caffeine is not considered a nutraceutical, and excess intakes can be potentially harmful.[76]

The major class of nutraceutical compounds found in coffee, tea, and cocoa is phenolics, but the types of phenolics are different. A serving of coffee was found to contain four times the antioxidant activity of a similar serving of tea or cocoa.[77] Coffee is rich in chlorogenic acid, with robusta containing about 25% more than arabica. Depending on the method of preparation, a cup of coffee may contain 15 to 325 mg of chlorogenic acid.[78] Another phenolic in coffee is caffeic acid, which has been reported to protect against oxidation in cell culture and animal models.[79,80] In moderation, coffee, like all foods including alcohol, can have beneficial properties because of its nutraceutical content.

Green tea and black tea have been found in epidemiologic studies to be associated with decreased incidences of CHD, cancer, and other diseases.[81-85] The phenolics in teas, which act as antioxidants, are considered to be the active ingredients conferring protection against these diseases. Green tea contains catechins (phenolics, class flavanols in Table 3.2), (–)-epicatechin (EC), (–)-epicatechin gallate (ECG), (–)-epigallocatechin (EGC), and (–)-epigallocatechin gallate (ECGC). In the black tea fermentation process, these catechins are oxidized and dimerized to form theaflavins.[86] The antioxidant potency of theaflavins was found to be similar to that of catechins.[86] Using a different method to measure antioxidant potency of phenolics than that used with coffee,[77] it was found that those in green tea and black tea were more effective in neutralizing free radicals compared to those in 33 commonly consumed fruits and vegetables.[88] Tea is often referred to as a beverage for relaxation or restoration. Green tea contains L-theanine, a unique amino acid, which is considered the compound associated with the restorative quality of tea.[89]

Cocoa contains phenolic compounds similar to those in tea, catechins and epicatechins, but differences exist. The complexes or polymeric forms of the catechins and epicatechins in cocoa (proanthocyanidins, procyanidins,

TABLE 3.2
Nine Basic Classes of Nutraceuticals (Functional Food Ingredients) in Foods with Examples[a]

I. Additives
    A. Fat substitutes
        1. Olestra
        2. Salatrim
    B. Medium-chain triglycerides
    C. Plant stanols
        1. Sitostanol
    D. Sugar alcohols or polyols
        1. Maltitol
        2. Mannitol
        3. Sorbitol
        4. Xylitol
    E. Sugar substitutes
        1. Aspartame
        2. Saccharin
        3. Sucralose

II. Botanicals
    A. Aloe
    B. Asian ginseng
    C. Black cohosh
    D. Echinacea
    E. Feverfew
    F. Ginger
    G. Ginkgo biloba
    H. Goldenseal
    I. Hawthorn plant
    J. Kava-kava
    K. Licorice root
    L. Milk primrose
    M. Milk thistle
    N. Peppermint oil
    O. Saw palmetto
    P. Siberian ginseng
    Q. St. John's wort
    R. Valerian root

III. Carbohydrates
    A. Dietary fiber — traditional sources from fruits, vegetables, and whole grains
    B. Fagopyritols (buckwheat)
        1. A1 (*O*-alpha-D-galactopyranosyl-(1–3)-D-chiro-inositol)
        2. B1 (*O*-alpha-D-galactopyranosyl-(1–2)-D-chiro-inositol)

(continued)

TABLE 3.2 (CONTINUED)
Nine Basic Classes of Nutraceuticals (Functional Food Ingredients) in Foods with Examples[a]

- C. Isolated sources of dietary fiber
    1. Cellulose
    2. Hemicelluloses
    3. Pectins
    4. Fructans
    5. β-Glucan
    6. Psyllium
- D. Modified and chemically produced sources of dietary fiber components
    1. Polydextrose
    2. Resistant maltodextrin; fibersol-2
    3. Methylcellulose
- E. Mucilages and gums
    1. Agar
    2. Carrageenan
    3. Gum arabic
    4. Gum tragacanth
    5. Locust bean gum
    6. Xanthan
- F. Nonabsorbable and/or nondigestible mono- and disaccharides
    1. D-Tagatose (also considered a prebiotic)
    2. Cellubiose
- G. Prebiotics, trisaccharides and larger (DP Π 3)
    1. Fructooligosaccharides (FOS)
    2. Galactooligosaccharides (GOS)
    3. Fructans
- H. Resistant starches

IV. Elements
- A. Boron
- B. Chromium
- C. Vanadium
- D. Lithium

V. Lipids
- A. Simple lipids — fatty acids
    1. Conjugated linoleic acid (zoochemical)
    2. Linolenic acid
    3. Eicosapentaenoic acid (C20:5 ω-3)
    4. Docosahexaenoic acid (C22:6 ω-3)
- B. Complex lipids
    1. Sphingolipids
        a. Ceramides

TABLE 3.2 (CONTINUED)
Nine Basic Classes of Nutraceuticals (Functional Food Ingredients) in Foods with Examples[a]

        b. Sphingomyelins
        c. Cerebrosides
        d. Gangliosides
    2. Phospholipids
        a. Phosphatidyl choline
  C. Structured lipids (see Additives; co-listed)
    1. Olestra
    2. Salatrim
  D. Terpenes (based on isoprene units)
    1. Monoterpenes (2 isoprene units)
        a. D-Limonene
        b. Pinene
        c. Eucalyptol
        d. Perilillic alcohol
    2. Triterpenes (6 isoprene units)
        a. Plant sterols (phytosterols) and stanols
            1). β-Sitosterol
            2). Campesterol
            3). Stigmasterol
            4). Sitostanol (used as ester)
        b. Saponins (steroid glycosides)
    3. Tetraterpenes (8 isoprene units)
        a. Carotenoids
            1). β-Carotene
            2). Lycopene
        b. Xanthophylls (tetraterpenoids)
            1). Lutein
            2). Zeaxanthin
  E. Quinones (phenolics) with isoprene side chains
    1. Tocotrienols: α, β, γ, and δ
    2. Others: Vitamin E (tocopherol), vitamin K (phylloquinone and menaquinone), and vitamin Q (ubiquinone)
VI. Nitrogen compounds
  A. Protein (zoochemical)
    1. Animal protein (meat factor effect)
  B. Peptides, whey protein hydrolysates
  C. Amino acids
    1. L-Arginine
  D. Capsaicinoids
    1. Capsaicin

(continued)

TABLE 3.2 (CONTINUED)
Nine Basic Classes of Nutraceuticals (Functional Food Ingredients) in Foods with Examples[a]

VII. Phenolics
    A. Simple phenols C_6
        1. Catechol
        2. Hydroquinone
    B. Benzoquinones C_6
        1. 2, 6-Dimethyoxybenzoquinone
    C. Phenolic acids C_6–C_1
        1. Salicylic acid
    D. Acetophenones C_6–C_1
        1. 3-Acetyl-6-methoxybenzaldehyde
    E. Phenylacetic acids C_6–C_2
        1. *p*-Hydroxyphenylacetic acid
    F. Hydroxycinnamic acids C_6–C_3
        1. Caffeic acid
        2. Chlorogenic acid
        3. *p*-Coumaric acid
        4. Ferulic acid
    G. Phenylpropenes C_6–C_3
        1. Eugenol
        2. Myristicin
    H. Coumarins (C) and isocoumarins (I) C_6–C_3
        1. Aesculetin (C)
        2. Umbelliferone (C)
        3. Berfenin (I)
    I. Chromones C_6–C_3
        1. Eugenin
    J. Naftoquinones C_6–C_4
        1. Juglone
        2. Plumbagin
    K. Xanthones C_6–C_1–C_6
        1. Mangiferin
    L. Stilbenes C_6–C_2–C_6
        1. Lunularic
        2. Piceid
        3. Resveratrol, *cis* and *trans*
    M. Anthraquinones C_6–C_2–C_6
        1. Emodin
    N. Flavonoids C_6–C_3–C_6
        1. Chalcones

TABLE 3.2 (CONTINUED)
Nine Basic Classes of Nutraceuticals (Functional Food Ingredients) in Foods with Examples[a]

2. Dihydrochalcones
3. Aurones
4. Flavones: a. Apigenin; b. Baicalein; c. Chrysin; d. Diosmetin; e. Diosmin; f. Eupafolin; g. Eupatilin; h. Flavone; i. Hispidulin; j. Luteolin; k. Tangeretin; l. Techteochrysin
5. Flavonols: a. Fisetin; b. Galangin; c. Kaempferide; d. Kaempferol; e. Morin; f. Myricetin; g. Myricitin; h. Quercetin; i. Quercetrin; j. Rhamnetin; k. Robinin; l. Rutin; m. Spirenoside
6. Dihydroflavonol
7. Flavanones: a. Eriocitrin; b. Eriodictyol; c. Hesperidin; d. Isosakuranetin; e. Likvirtin; f. Liquiritigenin; g. Liquirtin; h. Naringenin; i. Naringin; j. Neohesperidin; k. Pinocembrin; l. Poncirin; m. Silybin; n. Tangeritin; o. Taxifolin
8. Flavanols: a. Catechin; b. Epicatechin; c. Epicatechin gallate; Epigalliocatechin gallate; e. Epigallocatechin; f. Flavan; g. Gallic acid
9. Flavandiol or leucoanthocyanidin or Flavanolols: a. Pinobanksin; b. Silibinin; c. Silymarin; d. Taxifolin
10. Anthocyanidin: a. Apigenidin; b. Cyanidin; c. Delphinidin; d. Malvidin; e. Pelargonidin; f. Peonidin; g. Petunidin
11. Isoflavonoids
    a. Isoflavones: i. Biochanin A; ii. Daidzein; iii. Formononctin; iv. Genistein (genistin in glycoside form); v. Glycitein; vi. Pratensin
    b. Coumestans: i. Coumestrol; ii. 4'-O-methyl-coumestrol
12. Biflavonoids
    a. Amentoflavone
13. Proanthocyanidins or condensed tannins

O. Lignans (L) and Neolignans (N) $(C_6–C_3)_2$
1. Matairesinol (L)
2. Pinoresinol (L)
3. Secoisolaricirasinol (L)
4. Eusiderin (N)

P. Lignins $(C_6–C_3)_n$

VIII. Probiotics
   A. Bacteria
   1. *Lactobacillus* sp.
   2. *Bifidobacterium* sp.
   3. *Escherichia coli*
   4. *Streptococcus* sp.
   5. *Enterococcus* sp.

(continued)

TABLE 3.2 (CONTINUED)
Nine Basic Classes of Nutraceuticals (Functional Food Ingredients) in Foods with Examples[a]

    6. *Bacteroides* sp.
    7. *Bacillus* sp.
    8. *Propionibacterium* sp.
  B. Yeast
    1. *Saccharomyces boulardii*
IX. Sulfur compounds — Organosulfur compounds — Biothiols
  A. Glucosinolates in *Brassica* plant family (approximately 50 primary compounds)
    1. Glucoraphanin produces sulforaphane (isothiocyanate; R–SCN)
    2. Glucobrassicin produces indole-3-carbinol (does not contain sulfur)
  B. Glutathione
  C. Lipoic acid
  D. *S*-alk(en)yl-L-cysteine sulfoxides; one is alliin in garlic and in other *Allium* family plants. Hydrolysis of alliin by allinase yields allicin and upon rearrangement can yield diallyl disulfide (DADS), considered to be one of the more active sulfur ingredients in garlic and onions.

[a] Among the nine classes, approximately 200 examples of individual nutraceuticals are presented.

and prodelphinidins) appear to be more abundant in comparison to other beverages containing phenolic compounds.[90–93] In a human study, which used a randomized, two-period crossover design, 23 subjects were fed diets containing 22 g cocoa powder and 16 g dark chocolate or diets without the cocoa and dark chocolate.[94] The diet with cocoa and dark chocolate provided approximately 446 mg of procyanidins per day. The total phenolic content and distribution of all phenolic compounds in the diets was not reported, but the absence of these data does not diminish the value of the study. In summary, the results of this study showed that the phenolic content in the blood of individuals eating the cocoa/chocolate diet increased. This was interpreted as increasing the antioxidant capacity of the blood and slowing the potential for circulating low-density lipoproteins (LDL) to oxidize. Additionally, a significant increase in the level of high-density lipoproteins (HDL), 4% compared to the control group, was observed. These observations support the idea that phenolics can reduce the incidence of CHD, but the effect would be a long-term one. Although not the definitive or final study, this research describes the potential benefits of phenolics, specifically the phenolics provided by cocoa and chocolate. As a point of quantitative comparison, the authors mentioned the reference by Arts et al.[95] which reported that dark chocolate contains 0.535 mg catechins per gram compared to 139 mg of catechins per liter of tea. Techniques for the proper identification and measurement of phenolics in foods are advancing[96,97] but remain a challenge when attempts to determine the bioavailability and efficacy of phenolics are also being made.[98,99]

Coffee, tea and cocoa are rich in phenolics. These are unique nutraceutical beverages that supply a mixture of phenolics, which primarily act as antioxidants. The evidence suggests that these beverages, when consumed in moderation, are providing protection against oxidative damage in the body. These benefits appear to be most advantageous for disease prevention when viewed on a long-term basis.

Juices are defined and perceived to be the whole and undiluted liquid extracts of fruits, vegetables, and possibly any other plant food. All juices are excellent candidates for nutraceuticals. In their book, *Economic Botany*,[100] Simpson and Ogorzaly list approximately 100 edible fruits and an equal number of edible vegetables commonly consumed throughout the world. Each one of these fruits and vegetables represents a source of nutraceuticals, and many different nutraceuticals can be found within each fruit or vegetable. The consumption and diversity of a variety of juices, and mixtures of juices made from fruit, vegetables, and mixtures of fruit and vegetables are huge. Maybe more importantly, the subsequent distribution, consumer acceptance, and consumption are hampered only by limited marketing resources and ingenuity. Citrus products provide popular fruit juices. Citrus fruits (lemon, lime, orange, grapefruit, and tangerine) contain a number of different nutraceuticals, which include the monoterpene D-limonene, a variety of phenolic compounds in the flavonoid subclass (the flavanones hesperetin and naringenin and the flavones tangeretin and nobiletin), and to a lesser degree, compared to other fruits and vegetables, carotenoids.[101] The monoterpenes (D-limonene) in citrus are the essential oils giving the peels of these fruits their distinctive fragrance.

D-Limonene, when administered to rodents, can suppress tumorigenesis caused by many different carcinogens and procarcinogens; these results have been extensively reviewed.[102,103] This monoterpene has also been found to suppress implanted tumors in a variety of rodents,[103] and when orally administered, has been shown to help stabilize patients with breast and colon cancer.[104] The major flavanones in orange and grapefruit juices are hesperetin and naringenin, respectively. Although these compounds are known to contribute to the bitter taste in these fruits, they have been shown to have anticancer activity in human breast cancer cells grown in culture.[105] The citrus industry is faced with the conundrum of knowing that citrus fruits may contain valuable nutraceuticals such as flavonoids, but at the same time, knowing that these compounds contribute much to the bitter taste of the fruit and rejection by the consumer.[37] As the fruit juice industries reduce the level of bitter-tasting compounds in their products, they might want to think about reducing the sugar content of their products to better fit the lower calorie needs of the consumer. Likewise, the dairy industries have provided products with fewer calories by reducing the fat.

Epidemiological studies continue to show that cancer risks and the incidence of CHD are inversely related to the consumption of green and yellow vegetables,[106] fruits,[107] and tomatoes,[108,109] which are technically fruits. The two common classes of nutraceuticals (Table 3.2) in all these plant foods are carbohydrates, which provide dietary fiber, and lipids, which include the carotenoids and xanthophylls. While the increased consumption of dietary fiber appears to help prevent CHD,[110,111] neither the benefits of dietary fiber nor the role it plays in preventing cancer has been demonstrated in experimental studies.[112,113] It may take long-term dietary fiber intakes much higher than the National Academy of Sciences' recommended 25 g per day for women and 35 g per day for men[110] for the positive effects of dietary fiber on the incidence of cancer to be seen. The availability of a greater variety of juice products containing endogenous dietary fiber and added dietary fiber could help consumers achieve higher dietary fiber intakes.

While the benefits of dietary fiber for cancer prevention are controversial, it appears that the tetraterpenes (carotenoids and xanthophylls) in vegetables and fruits provide the protection needed to prevent this disease in its many different forms. Tetraterpenes appear to distribute themselves in many locations throughout the body. Although the exact mechanism(s) of prevention are not absolutely proven, it is speculated that carotenoids act through antioxidant activity.[114] Much of the color of the citrus fruits can be attributed to tetraterpenes, which include primarily various levels of β-carotene, lycopene, zeaxanthin, cryptoxanthin, and other tetraterpenes to a lesser degree. Although the total carotenoid levels in citrus juices may average less than 0.25 mg per 100 g serving,[101] routine consumption can add to sustained intake of these antioxidants to help protect the body against oxidative damage. Among the tetraterpenes that can be provided in fruit and vegetable juices, β-carotene, lycopene, lutein, and zeaxanthin (the dihdroxycarotenoid isomer of lutein) have received most of the attention, both individually and collectively. These tetraterpenes have been associated with cancer prevention as well as prevention of CHD, cataracts, and age-related macular degeneration (ARMD).[106–109] Carotenoids appear to have a wide range of benefits in health promotion and disease prevention.

β-Carotene, compared to all other carotenoids, is often cited as the most potent quencher of singlet oxygen radicals.[115] Although scientific debate continues about whether carotenoids, xanthophylls, or vitamin E are the most effective antioxidants in the body, β-carotene is possibly the most abundant carotenoid in the diet.[101] Carrots, with values that range from 4 to 8 mg of β-carotene per 100 g, are among the plant foods having the highest level of β-carotene.[101] While diets high in β-carotene appear beneficial, numerous trials of β-carotene supplementation have not shown positive effects on prevention or treatment of various forms of cancer. In fact, in clinical studies in which the diets of individuals were supplemented with β-carotene, the studies were terminated[116] or had negative results.[117]

Lycopene, specifically as provided in tomato products, has received positive attention in terms of the prevention of prostate cancer.[108,109] Tomatoes can provide as much as 10 mg of lycopene per 100 g,[101] the highest level of any food. Additional sources of lycopene from fruits are red (blood) oranges, watermelon (4.8 mg/100 g) and pink grapefruit (1.5 mg/100 g).[101] Lutein and zeaxanthin have also received attention in terms of cancer and CHD prevention, but they also appear to be beneficial in the prevention of cataracts and ARMD.[50,51] The health benefits of carotenoids consumed in food appear to be higher than when ingested as supplements,[116,117] and fruit and vegetable beverages can serve as excellent delivery systems.

Other juices that may have health benefits include cranberry and blueberry, which are rich in the phenolic compounds proanthocyanidins. Although the anthocyanidins in these juices give them their color, it is the proanthocyanidins that are being investigated for their ability to prevent urinary tract infections (UTIs).[118]

Blends of juices offer a variety of nutraceuticals. Possibly the best example on the food shelf is Campbell's V8® Juice, which contains tomato and seven other vegetables, including carrot, celery, beets, parsley, lettuce, watercress, and spinach. Although the exact amounts of tomato and the other seven vegetables contained in V8 are not publicly known, V8 is a juice with a broad array of nutraceuticals, predominately the tetraterpenes in the lipid class of nutraceuticals. Of the nine classes of nutraceuticals (Table 3.2), V8 Juice contains six classes — all except food additives, botanicals, and probiotics. It is acknowledged that many of the nutraceuticals in these six classes are present in minor to trace amounts, but V8 offers a variety of nutraceuticals in a serving. A modified V8 Juice could contain all classes. The following is a brief summary of the more prevalent nutraceuticals provided in V8: lycopene from tomatoes, β-carotene from carrots, D-limonene from celery, betalains (betacyanin and betaxanthin as potential antioxidants) from beets, myricetin and tannins from parsley, zeaxanthin and lutein from lettuce, and glucosinolates (converted to isothiocyanates) from watercress, which is in the family Cruciferae and genus *Brassica*. And, as reported on the V8 label, an 8 oz. serving of V8 provides 2 g of dietary fiber.

Soda is defined as a carbonated beverage containing high-fructose corn syrup, sucrose, or artificial sweeteners; phosphoric acid; flavors; and may contain caramel color. Soda, which is also referred to as soft drinks and pop, contains no alcohol. From personal experiences, a soda gives a feeling of well being through relaxation. The psychological value or feeling of well being derived from these beverages when consumed in moderation cannot be discounted. Although soda is maligned for its high sugar or high-fructose corn syrup content as a cause of increased obesity in the U.S., it can have redeeming qualities. When sugar substitutes (additives, Table 3.1) are used, these beverages offer tasty thirst-quenching and satisfying fluids, which are especially beneficial to individuals watching their weight or managing blood sugar levels associated with diabetes. Ideas to fortify soda with nutrients and nutraceuticals have never materialized.

Drinks are possibly the broadest category of potential nutraceutical beverages. Again, the multitude of combinations of water, juice, and/or nutraceutical supplements to make a drink are unlimited. Sport and stimulator beverages are variations of drinks, but listed separately. Water becomes a type of drink with added juice, vitamin C, St. John's wort, etc. Drinks may include beverages that are basically water, too, but with added flavoring(s) and coloring agents. Kool-Aid®, introduced in 1927, is best known as a "drink." Current packages contain vitamin C (10% of the Daily Value [DV]), with no other nutrient, except that one serving provides 5 mg of sodium; a minimal amount. Is Kool-Aid a nutraceutical beverage? To the authors, the answer is a qualified yes, because of its vitamin C content, and when made with sugar substitutes, it is a practical, thirst-quenching beverage with few calories and almost no sodium — an advantage for individuals who need to control their sodium intakes. A quick tour of a supermarket provides a wide variety of liquid or powdered drink formulas containing, but not limited to, bioflavonoids, standardized herbs, vegetable extracts, cell pigments, whole foods, plant enzymes, soy protein, spirulina, etc. The number of ingredients that can be formulated to produce nutraceutical drinks is unlimited.

Sport beverages have a special purpose and for that reason may not be thought of as first-line nutraceutical beverages. Sport beverages are often called sports drinks, carbohydrate–electrolyte beverages, electrolyte replacement drinks, or isotonic drinks. They are intended for use by athletes, but also by workers who perform strenuous activity for an hour or longer. The following is a quote from the American College of Sports Medicine:[119] "During exercise lasting less than 1 h, there is little evidence of physiological or physical performance difference between consuming a carbohydrate–electrolyte drink and plain water." However, because of the moderate levels of readily available carbohydrate and electrolytes that more rapidly hydrate the individual, sport beverages are commonly prescribed to replace electrolytes in children with diarrhea. Thus, this is an example of a sport beverage becoming a nutraceutical beverage. Furthermore, needed or not, they are a popular beverage among active people. Usually they have about one-half the calories of fruit juices (one-half the sugar) and have mixtures of carbohydrates and electrolytes not typically found in juices, juice drinks, or water.

Some may contain herbs or supplements, and these ingredients are said to enhance energy, endurance, or weight loss. However, these supplements may actually distract from the ability of the carbohydrates and electrolytes in balanced sport drinks to achieve optimum absorption, performance, and endurance. Too much of some supposedly good things (protein, vitamins, and other minerals) may actually slow absorption and hydration of the body. While caffeine is often described as a stimulant that provides stamina and endurance, athletes are not advised to consume caffeine for performance. Although sport beverages were originally designed for very competitive amateur and professional athletes, they are widely accepted as energy boosters for the average person.

Gatorade® has possibly become the quintessential sport beverage since it was commercially introduced in 1967 and promoted for electrolyte replacement. Sport beverages are basically designed to help avoid fatigue and improve endurance, while preventing muscle cramps, usually associated with dehydration. A standard of identity for sport beverages has not been established. Researchers continue to investigate the optimal levels of carbohydrates and electrolytes in water volume.[120] At this point, research suggests that the optimum level of carbohydrates (glucose, sucrose, and/or soluble multi-dextrins) is 6 to 7%, with smaller amounts of sodium, potassium, chloride, and phosphate; and the optimum sodium levels cited are 100 to 110 mg per 100 ml or approximately 45 mmol/l.[121] The osmolality (i.e., the number of particles in solution) of sport beverages ranges from 208 to 380 mosmol/kg water.[121] The carbohydrates, if provided with correct levels of electrolytes, provide quick energy because of their almost immediate absorption. Although in periods of moderate activity and for periods of time less than one hour, water and sport drinks are equally absorbed, since the rate of absorption is the same, the addition of carbohydrates and primarily electrolytes helps completely hydrate the body and its cells in shortened periods of exercise or work. As mentioned, the real value of a sport beverage is to the person exercising or participating in a strenuous activity for at least an hour. Water appears to be effective for quenching thirst and hydration during short periods of strenuous exercise. The appropriate consumption of a sport beverage with readily available energy and electrolytes appears to provide a competitive edge.[122,123]

Adequate fluid intake is recommended before athletic events, and this need can be met with water. However, during and after strenuous physical activity, hypotonic or isotonic sport drinks are recommended.[124] The formulation of sport beverages has been reviewed in regard to the carbohydrate and electrolyte levels.[121] The terms hypotonic, hypertonic, and isotonic are frequently associated with sport beverages, and these descriptors are best defined in relationship to osmolality in the human body. The normal osmolality of the intestinal lumen in a fasted individual ranges from 270 to 290 mosmol/kg; this is also the isotonic value of human serum. Absorption from the intestine is optimal at or near this osmolality. Water or hypotonic mineral waters have osmolalities that range from 5 to 15 mosmol/kg, which are very low. When consuming these hypotonic beverages, it takes time to achieve isotonicity in the intestinal lumen; electrolytes must come into the intestine, and thus absorption is slowed. A hypotonic sport beverage has an osmolality slightly below 270 mosmol/kg; water is not considered a hypotonic sport beverage. When an isotonic sport beverage is consumed, absorption of water and energy (carbohydrates) is optimal. The osmolalities of orange juice and Coca-Cola Classic® are 663 and 700 mosmol/kg water, respectively.[121] When these beverages are consumed during or after prolonged physical activity, water must be transferred from the circulatory system into the intestine to achieve isotonicity. This takes extra time, and the sugars and electrolytes move farther down the intestine and, in effect, make the hypotonic solution ineffective for promotion of rapid hydration. For optimum hydration during and after physical activity, sport beverages that are slightly hypotonic or isotonic appear to be the most effective. Inexpensive and readily available hypotonic and isotonic sport drinks can be made by diluting orange juice with three parts water and adding 1 g of salt and diluting orange juice with one part water and adding 1 g of salt, respectively.

Stimulator beverages are unique because they attempt to imply that the beverages provide "energy" as a nutrient (carbohydrate), when in fact they are providing stimulants for alertness. Stimulator beverages are not to be mistakenly considered sport drinks for performance. Furthermore, stimulator beverages should not fall into the category of nutraceutical beverages. They are briefly discussed because of their association with energy and claimed contribution to endurance. Often these drinks are referred to as energy-providing or high-energy drinks. In marketing and advertising, the common tag line for these products is "Stimulation for Body and Mind." The term stimulator beverage was chosen for this review rather than the more common terms of energy drink or performance enhancer. Caffeine and taurine are the primary active ingredients in these beverages. These drinks usually contain sugar and various water-soluble vitamins (B vitamins), which can include riboflavin, niacin, and vitamins B_6 and B_{12}. These vitamins are cited in basic nutrition texts as being necessary for the metabolism of carbohydrates for energy. Thus the association: because the drink contains these vitamins, energy is made available faster. Some stimulator beverages include guarana, which is a natural source of caffeine. The caffeine levels found in stimulator beverages might be higher than the levels found in colas and other carbonated soft drinks. Common sodas or soft drinks contain approximately 50 mg of caffeine per 12 oz. Moderate caffeine consumption is considered to be about 200–300 mg per day, the amount in 2 to 3 cups (8 oz. each) of coffee. The levels of taurine added to these beverages are not reported, and reporting is not required by FDA regulations. The list of purported physiological or health benefits of taurine is among the most numerous of any amino acid or nutraceutical.[125] The inclusion of taurine in stimulator drinks is considered to be justified possibly because this amino acid is claimed to have detoxifying effects. But more important, taurine is claimed to be inotropic (able to influence the force of muscular contractility), an effect that helps regulate the highs and lows of blood calcium in the heart. The theory is that if caffeine can stimulate, taurine will help the heart to uniformly operate (pump) during these periods of stimulation. However, blood calcium levels are highly regulated by parathyroid hormone and do not change in a healthy person. Taurine is the second most abundant amino acid in the blood, and although it is not directly made in the body, the body easily converts cysteine to taurine. Taurine is also ubiquitous in all foods. These stimulator beverages are not advisable during or after strenuous exercise because of their high caffeine content and high osmolality (see sports beverages). Some of the most successful stimulator beverages on the market include Red Bull®, Lipovitan®, Solstis®, Red Alert®, Vialize®, Life Plan®, and Red Devil®.

Wellness (nutraceutical) beverages could be a composite of the six categories of beverages listed in Figure 3.1, except stimulator beverages. Since all foods are functional foods, including water, all types of beverages could be generally described as wellness beverages that can contribute to health. The challenge food scientists, nutritionists, and health professionals have today is to match a functional food with consumer demand for a health product. Then, the active ingredient(s) must be identified, and clinical studies must be completed to show efficacy or a cause-and-effect relationship. Nutraceutical-based foods have two additional challenges for their development and successful marketing based on scientific studies: how best to deliver the active ingredient in a nutraceutical beverage and how to provide it in adequate amounts. Lycopene is an example. Whole tomato juice contains an average of 10 mg of lycopene per 8 oz. serving. If it is shown that larger amounts of lycopene would be beneficial, how should this best be accomplished? Should individuals consume lycopene concentrates in liquid or tablet forms? Should lycopene be isolated and added to increase levels in the tomato juice, or should plant breeding or genetic engineering be used to substantially increase the amount of lycopene in the tomato? A broad summary of scientific evidence suggests that the delivery of nutraceuticals in foods is more beneficial than delivery in isolated form because of higher bioavailability. Yes, there are exceptions to this statement. The maximum amount of any nutraceutical to be consumed over short or extended periods of time for

TABLE 3.3
Classes of Biochemical, Physiological, or Molecular Mechanisms for Proposed Action of Nutraceuticals Possibly Contributing to Improved Health[a]

I. Antioxidants
   A. Lipids: tetraterpenes
      1. Lycopene in tomatoes (165)
      2. Lutein and zeazanthin in spinach (165)
   B. Phenolics
      1. Catechins:(−)-epigallocatechin and (−)-epigallocatechin-3-gallate in green tea (88)
      2. Resveratrol in wine and peanuts (43)
II. Antiinflammatory
   A. Lipids: fatty acids
      1. Eicosapentaenoic acid in fish (C20:5 ω-3) (195)
      2. Docosahexaenoic acid in fish (C22:6 ω-3) (195)
III. Antibacterial
   A. Carbohydrates
      1. Galactooligosaccharides (GOS) in human milk (63)
   B. Phenolics
      1. Genistein and daidzein in soy (196)
      2. Proanthocyanidins in cranberries and blueberries (118)
   C. Sulfur compounds
      1. Diallyl disulfide (DADS) in garlic and onions (197)
IV. Effects on signaling (membrane and messenger)
   A. Lipids: fatty acids
      1. Eicosapentaenoic acid in fish (C20:5 ω-3) (178)
      2. Docosahexaenoic acid in fish (C22:6 ω-3) (178)
   B. Phenolics
      1. Genistein in soy (198)
V. Gene expression
   A. Lipids
      1. Carotenoids (199)
   B. Probiotics; intestinal bacteria (200)
VI. Cell kinetics (differentiation, propagation, and apoptosis)
   A. Carbohydrates: dietary fiber (201)
VII. Hormonal actions
   A. Lipids
      1. Eicosapentaenoic acid in fish (C20:5 ω-3) (142)
   B. Phenolics
      1. Genistein and daidzein in soy (49)
      2. Lignans in flax and rye (153,154)

TABLE 3.3 (CONTINUED)
Classes of Biochemical, Physiological, or Molecular Mechanisms for Proposed Action of Nutraceuticals Possibly Contributing to Improved Health[a]

VIII. Energy for intestinal bacteria
    A. Carbohydrates
        1. Prebiotics (126)
        2. Resistant starch (173)

IX. Immune stimulators
    A. Carbohydrates
        1. Chitin in fungi (180)
    B. Lipids
        1. Eicosapentaenoic acid in fish (C20:5 ω-3) (178)
        2. Docosahexaenoic acid in fish (C22:6 ω-3) (178)
    C. Nitrogen compounds
        1. L-Arginine (202)
    D. Probiotics (203)

X. Intestinal bulk
    A. Carbohydrates
        1. Wheat bran (181)
        2. Psyllium (182)

XI. Intestinal absorption (factors affecting)
    A. Additives
        1. Calcium carbonate–citric acid–malic acid (CCM) (7)
    B. Carbohydrates
        1. β-Glucan in oats (204)
    C. Lipids
        1. β-Sitosterol and sitostanol in plant oils and hydrogenated forms, respectively (186)
    D. Nitrogen compounds
        1. Animal protein; "meat factor" effect (184)

XII. Enzyme activity (inhibitors/activators)
    A. Nitrogen compounds
        1. Milk peptides (148)
    B. Sulfur compounds
        1. Isothiocyanates in cruciferous vegetables (188)
        2. Genistein and daidzein in soy (198)

[a] Examples of foods and nutraceuticals are presented that may exert one or more mechanisms of action. Numbers in parentheses pertain to references for mechanism of action for a representative nutraceutical. This is not a complete list and is intended to summarize and complement information provided in the text of this chapter.

TABLE 3.4
List of Suggested Criteria to Establish Identity, Chemical and Physical Properties, Bodily Functions, and Efficacy of a Nutraceutical

1. Common and scientific names, chemical formula, molecular structure (to include complexes and compound molecules)
2. Chemical reactivity and stability (Is there a difference between the nutraceutical naturally occurring in foods versus that isolated and in a purified form?)
3. Solubility
4. Common method(s) of isolation, measurement and purification (consider complexes and compound molecules)
5. Food sources
6. Amount in foods
7. Location in foods
8. Function or purpose in foods
9. Function or purpose in food quality and acceptance
10. Mammalian absorption/bioavailability
11. Location and storage in animals and humans
12. Functions/mechanisms/biochemistry/physiology/molecular actions in the animal and human body
13. Can the amount be determined and or estimated in the animal or human body? How?
14. Can the effects of ingestion and retention be determined?
    a. Is there a response? What are the endpoints?
    b. How?
    c. Is there a biomarker for nutraceutical or function?
15. Diseases or disorders associated with
16. Amount needed for chronic or acute response
    a. How much is needed to aid in a bodily function?
    b. How much is needed and for how long to treat a disease or disorder?
    c. Is nutraceutical prophylactic, therapeutic, or both?
17. Interacts with
    a. Adjuvant to nutrients or other nutraceuticals
    b. Negative interactions
18. Upper Limits (UL) — toxicity (LD_{50})
19. References for items 1–18

optimum health remains to be determined. Again, these are important challenges. For persons interested in developing a nutraceutical beverage or food, a checklist of information, designed to help you acquire a range of knowledge from basic understanding to advantageous marketing, is provided in Table 3.4. This information is similar to the information presented for a food ingredient/nutraceutical petition for FDA self-affirmed Generally Recognized as Safe (GRAS) status or for a health claim.

CLASSES OF NUTRACEUTICALS

In previous sections of this review, attention was given to natural foods as beverages. As in any physical science discipline, the taxonomy of nutraceuticals may be helpful. We elected to divide functional foods into nine chemical categories. An alternative would be to divide these chemicals, since most are organic, into classical groups based on their elemental composition and then their functional or reactive groups as described in basic and advanced organic chemistry texts. However, nutraceuticals are more easily comprehended when they are described first as food chemicals and later as individual organic compounds. Again, some of these compounds can be cross-listed among the nine classes. For example, carrageenan is a carbohydrate (hydrocolloid) and source of dietary fiber and a sulfur-containing compound. Medium-chain triglycerides can be described as food additives and lipids. Many functional food ingredients listed in Table 3.2 may not be mentioned here but warrant additional review.

Table 3.2 divides nutraceuticals into nine classes based on their manufacture or simple chemical composition and characteristics. One group, probiotics, is for living organisms, and does not consist of chemicals. Within each class, there are subclasses. Examples of some of the more common individual nutraceutical compounds, frequently mentioned in the scientific literature, are given in each of the nine classes, but this is not a complete list. Readers are encouraged to expand this list based on their knowledge of nutraceuticals.

The class of food *additives* refers to those food ingredients that have been isolated or specifically developed and manufactured and are now used in foods for health purposes. Some may exist in foods in small amounts. Examples of these additives include the polyols, sugar and fat substitutes, bulking agents and prebiotics (fructooligosaccharides, polydextrose, and resistant maltodextrin), and stanols. This list is not inclusive of all the food additives that have potential nutraceutical properties. Polydextrose, resistant maltodextrin (Fibersol-2®), and fructans can also be classified as carbohydrates and are used as sources of dietary fiber, especially in beverages, because of their high solubility. Polyols have wide application in candy and gums to help prevent dental caries. Polyols are carbohydrates but are not considered to be sources of dietary fiber. Obesity, which includes individuals with a Body Mass Index over 30, is now a serious worldwide health problem. Excess sugar and fat have often been targeted as the primary dietary factors leading to excess weight and obesity. The sugar substitutes (aspartame, saccharin, and sucralose) are not only effective for sugar (calorie) intake reduction, but foods and beverages containing these sugar substitutes provide an excellent alternative for diabetics managing their blood glucose levels.

Two examples of fat substitutes, olestra and salatrim, provide 0 or 5 kcal/g, respectively, compared to conventional fats/oils that have 9 kcal/g. Although these fat substitutes are effective in reducing the fat and calorie content of foods, consumer consumption of foods containing fat substitutes has waned. This decline can be traced, at least in part, to the lack of commentary by government agencies and other public health agencies, which could specifically indicate that fat consumption is a cause of weight gain. Another example of a nutraceutical listed as a food additive or manufactured functional food ingredient is the stanols produced by hydrogenation of plant sterols, which are designed to help lower blood cholesterol levels. Stanols interfere with the intestinal absorption of dietary cholesterol. The fat substitutes and stanols can also be classified as lipids. Except for use in soups, and liquid meal replacements and medical foods, these lipid compounds would have little use in beverages.

Botanicals were among the first functional food ingredients to seriously captivate the consumer. With claims that ranged from curing depression to enhancing memory, botanicals made consumers take notice. Some of the major botanicals consumed worldwide include echinacea for upper respiratory tract infections and stimulation of the immune system; St. John's wort for depression; ginkgo biloba for enhanced cognitive function; saw palmetto for the prevention of benign prostatic hyperplasia (BPH); and Asian ginseng for sedation. A more extensive list is provided in Table 3.5. Botanicals, as a class of functional food ingredients, are extremely popular, and clinical research designed to verify their safety and efficacy is ongoing. However, many of these studies have not provided unequivocal proof of the claimed efficacy of these botanicals. A serious concern with botanical sales is product purity and identification of the active ingredient(s). Food manufacturers are advised to proceed with caution when using botanicals in their products.

Two primary subcategories of *carbohydrates* are dietary fiber and prebiotics. The history of dietary fiber is long, and the proposed health benefits of dietary fiber may cover a broader spectrum of diseases and disorders than those of any other functional food ingredient. The U.S. National Academy of Sciences recently established the first Dietary Reference Intakes (DRI) for dietary fiber at 38 g/day for men and 25 g/day for women.[110] Median intakes of dietary fiber in the U.S. for men and women are 17 and 13 g/day, respectively.[110] The disparity between the recommended and actual intakes represents a great opportunity to find and use appropriate sources of dietary fiber in beverages.

Prebiotics are defined as "a nondigestible food ingredient that beneficially affects the host by selectively stimulating the growth and/or activity of one or a limited number of bacteria in the colon."[126] Intestinal health is associated with intake of dietary fiber and prebiotics.[127] One form of potential prebiotic

TABLE 3.5
Potential Benefits and Possible Adverse Effects Associated with the Consumption of Botanicals (Herbal Medicines)[a]

Botanical	Potential Benefits	Active Agent(s)	Site(s) of Action(s)	Complications or Side Effects
Aloe	Laxative, indigestion, sunburn	Aloin (phenolic)	Intestine, skin	Stomach discomfort and headache; avoid in pregnancy
Black cohosh	Prevent hot flashes and other menopausal symptoms	Fukinolic acid; triterpene glycosides (saponins); actein and cimicifugoside	Unknown; possibly estrogenic activity	
Echinacea	Enhance immune system; relieve common cold, flu, coughs	Echinosides, cafferic acid, ferulic acid (phenolics)	Immune system	Mild allergic reactions due to similarity to ragweed; immunosuppressant; limit use to 6 to 8 weeks
Ephedra[b]	*Use is strongly discouraged*			Irregular heartbeat; hypertension; risk of stroke
Evening primrose[c]	Relieve rheumatoid arthritis; premenstrual syndrome	γ-Linolenic acid	Moderate inflammation reduced by production of prostaglandin E_1 (PGE_1)	May increase the anticoagulant effect of drugs such as warfarin; do not use with anticonvulsant medication; gastrointestinal upset; nausea; loose stools; headache; seizure
Feverfew	Relieve migraine headaches	Sesquiterpene lactones such as parthenolide	Brain	
Ginger	Postoperative nausea; motion sickness	Gingerol and gingerdiols (volatile oils)	Intestine	

(continued)

TABLE 3.5 (CONTINUED)
Potential Benefits and Possible Adverse Effects Associated with the Consumption of Botanicals (Herbal Medicines)[a]

Botanical	Potential Benefits	Active Agent(s)	Site(s) of Action(s)	Complications or Side Effects
Ginkgo biloba[c]	Enhance memory and alleviate effects of dementia as in Alzheimer's disease	Ginkgo-flavone glycosides; terpene lactones	Blood thinner	Increased risk of bleeding when taken with other anticoagulant drugs (e.g., aspirin)
Ginseng[c]	Reduce fatigue and enhance metabolism during exercise; aphrodisiac	Ginsenosides and saponins		Hypoglycemic; increased risk of bleeding; can cause headache when used with antidepressants; caution with insulin
Hawthorn	Angina; cardiac insufficiency; hypertension	Procyanidins; saponins; cardiac active amines	Circulatory system	
Kava-kava[c,d]	Relieve anxiety; often used as an anticonvulsant; *use is strongly discouraged*	Kavalactones		Severe cardiac insufficiency; dose-related sedation; hypotension
Licorice root	Expectorant to treat coughs	Saponins; flavonoids		Can cause liver damage
Milk thistle	Liver disorders; patients infected with the human immunodeficiency virus (HIV); breastfeeding problems	Silibin; silidianin; silicristin	Liver	Nausea, diarrhea; allergic reactions in some people

Peppermint oil	Relief of irritable bowel syndrome; gastritis	Menthol	Intestine	
Saw palmetto[c]	Benign prostate hyperplasia	Fatty acids, sterols	Prostate gland; α-reductase inhibitor	
St. John's wort[c]	Alleviate mild to moderate depression, related anxiety, and insomnia	Hypericin, pseudohypericin		May reduce efficacy of conventional medications such as steroids
Valerian root[c]	Mild sedative; sleep aid	Valeportrites, valerenic acid, sesquiterpenes	Brain	Headaches

[a] See references (192) and (193).
[b] The use of ephedra is discouraged by many medical authorities. There has been no formal action by FDA.
[c] Specifically listed by the Government Accounting Office (GAO) as having potentially adverse side effects (194).
[d] The use of kava is discouraged by many medical authorities, and kava has been banned as unfit in the U.K. There has been no formal action by FDA.

found in cereals and legumes more often than other plant foods is resistant starch.[128] However, resistant starch does not appear suitable for use in beverages because of poor solubility. The use of resistant starch may not be effective in soups, liquid meal replacements, or medical foods. Heating these beverages would probably make the resistant starch susceptible to digestion in the small intestine. Two of the most common sources of prebiotics extensively investigated for a broad range of health properties are fructooligosaccharide (FOS), degree of polymerization (DP) = 3, and inulin, a fructan polymer that may range in size between 4 and 70 DP units.[129] Fructans with intermediate lengths (DP 4–10) and longer (DP 10–70), can have wide application in beverages. The soy polysaccharide obtained from soy cotyledon, a hemicellulose, can be described primarily as an insoluble dietary fiber but has excellent suspension in liquid meal replacements and medical foods. Polyols containing one and two saccharide units have wide application in beverages to help retard tooth decay;[130] however, excessive consumption will lead to diarrhea because of poor absorption and subsequent increased solute load in the large intestine.[131] A single-unit saccharide, D-tagatose, is not absorbed and is claimed to be a prebiotic for use in beverages.[132]

No single carbohydrate or potential source of dietary fiber has generated as much interest in the past 3 to 5 years as FOS. Commonly referred to as a prebiotic,[126] according to claims, a daily regime of 3 to 4 g of FOS per day is sufficient to alter the intestinal microflora, enhancing the lactic acid producing *Lactobacillus* sp. and *Bifidobacteria* sp., while reducing the levels of pathogenic bacteria.[133] More recently, reports are suggesting that the fermentation of FOS in the large intestine will significantly enhance the absorption of calcium from the colon.[134] If these data prove to be correct, this increase in the bioavailability of dietary calcium, which is not thought to be normally significantly absorbed in the large intestine, would have a significant impact on the prevention of osteoporosis. Research accomplishments in this area should be carefully monitored.

Ten *elements* have been established as essential. Calcium is possibly the most frequently cited element as a functional food ingredient for the prevention of osteoporosis. However, calcium represents an interesting situation. Calcium is an essential nutrient that is also marketed as a functional food ingredient. The introduction of the calcium-fortified orange juice beverage (60% juice) Citrus Hill (Procter & Gamble) in the early 1980s represented more then the addition of calcium to foods. For the first time, a complex of calcium (calcium carbonate–citric acid–malic acid) (CCM) was shown to give approximately 10% higher calcium bioavailability compared to other food sources of calcium and soluble calcium salts.[135] Normal calcium bioavailability in humans is approximately 50% of ingested amounts. Improved bioavailability of any nutrient or nutraceutical would be highly advantageous.

Other nonessential elements investigated for functional food properties include boron, possibly for embryo development and reproduction,[136] and lithium, for bipolar depressive illness.[137] Chromium has long been investigated for its possible role in restoring glucose tolerance.[138] Chromium picolinate is a popular mineral (element) supplement claimed to build muscle, reduce body fat, and allow people to lose weight without dieting, but there are no reliable data to support these claims.

Within the *lipid* class of nutraceuticals are subclasses. and within each subclass are many well-publicized nutraceuticals. These include fatty acids such as conjugated linoleic acid (CLA) (body fat reduction),[139] γ-linolenic acid (antiinflammatory),[140] eicosapentaenoic acid — C20:5 ω-3 (blood lipid lowering and reduced platelet aggregation),[141] and docosahexaenoic acid — C22:6 ω-3 (brain development and cognitive learning).[142] Examples of other lipids include the sphingolipids (control cancer cell growth),[143] the carotenoids β-carotene (antioxidant),[115] and lycopene (prostate cancer prevention),[144] and the xanthophylls lutein and zeaxanthin (cataracts and ARMD).[145] Two additional lipid nutraceuticals are the monoterpene D-limonene (tumor suppression),[102] found in orange peel, and the triterpene saponin (blood cholesterol reduction),[146] found in a variety of plants. The long-chain fatty acids, especially the omega-3 fatty acids, are prone to oxidation, and this makes their addition to beverages problematic.

Examples of *nitrogen compounds* that have functional food properties include animal protein, peptides, and some amino acids. All animal muscle proteins, compared to plant proteins, have the ability to enhance nonheme iron absorption, which is often referred to as the "meat factor" effect.[147] While there is no reference available for a liquid animal muscle protein beverage to enhance iron absorption, among the list of FOSHU ingredients is heme-Fe (Table 3.1), which has high iron bioavailability. Soy milk, and to a lesser degree, rice milk products are becoming popular beverages. Although soy, rice, or cow's milks will not enhance nonheme iron absorption, they provide other nutraceutical benefits. An increasing number of peptides, primarily obtained from the hydrolysis of milk proteins, are promoted for their antihypertensive effects;[148–150] see Table 3.1. L-Arginine is suggested to be cardioprotective because it serves as the precursor for nitric oxide, a known vasodilating substance; blood vessels are kept open and flexible.[151]

An estimated 8000 different *phenolic compounds* occur in nature. Although not all are common in foods, much remains to be learned about how these compounds may affect bodily processes and ultimately health. From a chemical perspective, these phenolics can be divided into 16 subclasses, with a majority of the biologically active compounds in the flavonoid class ($C_6C_3C_6$), which has 13 subclasses of approximately 4000 compounds (Table 3.2).[152] Genistein and daidzein in soy have been investigated extensively as phytoestrogens,[70] which might help prevent breast cancer. Another important phytoestrogen, also proposed to prevent breast cancer, is lignan (secoisolariciresinol-diglyceride), which is found in high amounts in flaxseed[153] and in lesser amounts in rye.[154] These plant lignans are converted by gut microflora to the active compounds enterodiol and enterolactone and then absorbed from the colon into the circulatory system. When activated in the intestine and absorbed into the bloodstream, these plant lignans are referred to as human lignans. They are being intensively investigated for their ability to moderate high estrogen levels, which are thought to be a major contributing factor in the progression of breast cancer.[153] Some phenolics prevent pathogenic bacteria from adhering to urinary tract membranes (proanthocyanidins in cranberries and blueberries).[118] Red grapes and especially red wines are high in anthocyanidins (flavonoids), which have antioxidant properties. Resveratrol (nonflavonoid; a stilbene) in grapes and wine (and peanuts) is reported to help protect circulating blood lipoproteins, specifically, from oxidation, by acting as an antioxidant.[43,44] But other phenolics in wine may be more effective.[45] Red wine is commonly associated with the term "French paradox" because of its high phenolic acid content. Scientific evidence continues to suggest that phenolics in wine will retard CHD. The term bioflavonoids does not refer to a class of phenolic, but rather is a collective term for the different phenolics found in citrus fruits that were reported to prevent capillary fragility. The collective group of flavanones in citrus fruits is referred to as bioflavonoids and has also been called vitamin P. These phenolics were associated with vitamin C because of their purported beneficial effects on capillary function.[155]

Probiotics are defined as "live microorganisms, which when administered in adequate amounts, confer a health benefit on the host."[156] Probiotics are used to prevent or treat intestinal disorders, such as bacterial infections, antibiotic-induced and rotavirus-caused diarrheas, and traveler's diarrhea.[157] Many probiotics appear to aid the lactose-intolerant individual by effectively hydrolyzing lactose when consumed with dairy products.[158] The consumption of probiotics appears to both complement the cells lining the intestine and displace potential pathogenic cells throughout the entire intestinal tract.[159] Probiotics also appear to enhance or prime the intestinal immune system.[160] Probiotics are occasionally prescribed and used as drugs, "living drugs," and are possibly the only example of a functional food also used as a drug.[161] Probiotics must be consumed on a regular basis for proposed effectiveness as they do not colonize with the intestinal microflora.[162] Like all of the nutraceuticals mentioned briefly in this review, the role played by probiotics in intestinal health maintenance and the value of prebiotics for intestinal microflora warrant further review. Probiotic beverages are very popular in Asia and Europe and are gaining in popularity in South America, but they do not have firm consumer acceptance in North America. This may be due to a lack of understanding of the concept of intestinal health, which is the cornerstone of probiotic consumption.

Sulfur compounds give many foods their characteristic aromas and tastes. Onions and garlic are rich in alliin (*S*-allyl-L-cysteine sulfoxide), which is odorless. Enzymatic conversion of alliin to allicin, which has odor, is followed by conversion to diallyl disulfide (DADS). Although DADS is thought to be the active component of garlic, conveying antimicrobial, anticarcinogenic, and blood cholesterol–lowering effects, other metabolites must also be considered active agents. Isothiocyanates (indole-3-carbinol, sulforaphane, and benzyl isothiocyanate) are derived from the hydrolysis of glucosinolates in plants of the family Cruciferae and genus *Brassica* (Brussels sprouts, broccoli, and cauliflower). Isothiocyanates are reported to induce liver enzyme systems (Phase I and Phase II) that upregulate carcinogen detoxification enzymes. An excellent review of sulfur compounds (glucosinolates and *S*-allyl-L-cysteine sulfoxide) and their nutraceutical properties is recommended.[163]

BIOCHEMICAL, PHYSIOLOGICAL, AND MOLECULAR ACTIONS OF NUTRACEUTICALS

Essential nutrients are associated with a host of biochemical, physiological, and molecular reactions in the human body. Essential nutrients are often referred to as the necessary dietary cofactors needed to make the biological system function. With sufficient scientific data, can it be said that nutraceuticals are also needed for optimum health? The science of nutrition is based on elucidating the biochemical, physiological, or molecular pathways, and the multitude of reactions that occur because of these pathways is immense.[164] Although nutraceuticals are not considered replacements for essential nutrients, they appear to complement many of the same reactions, which ultimately lead to the prevention and possible cure of chronic diseases. For example, vitamins E and C are considered antioxidants with antioxidant activity that is complemented by carotenoids and phenolics. Increased dietary calcium can be beneficial in the prevention of osteoporosis, but if prebiotics can be shown to enhance calcium bioavailability, it is possible that the incidence of osteoporosis could be greatly reduced.

Because not all the reactions influenced by essential nutrients are known, we are far from knowing all bodily effects of nutraceuticals. As Table 3.2 lists the major classes of nutraceuticals in foods and cites slightly more than 200 of the potential thousands of nutraceuticals in foods, Table 3.3 is an attempt to list some of the major classes of biochemical, physiological, and molecular actions of nutraceuticals in the body. Thirteen classes of actions are proposed and listed in Table 3.3; and while it might be naïve to consider only 13 classes, this is not considered a complete list but rather a list in progress. It is important to note that different classes of nutraceuticals can have common biological reactions. For example, both omega-3 fatty acids and probiotics are thought to enhance or stimulate the gut-associated lymphoid tissues (GALT), leading to a healthier or charged immune system. Tetraterpenes (carotenoids and xanthophylls) and phenolics are vastly different compounds chemically and physically, but both classes of nutraceuticals are thought to act as antioxidants through different chemical reactions or mechanisms. Phenolics are also unique in that different phenolic compounds can act as antioxidants, have hormone-like actions, and have antibacterial properties. Examples are given for nutraceuticals acting through different biological mechanisms of action in the following paragraphs.

Although nutraceuticals are often associated with the prevention of heart disease, cancer (suppressing the proliferation of cancer cells and suppressing the growth of implanted tumors), or diabetes, or cited as anticarcinogenic, antimutagenic, and hypolipoproteinemic, these are general statements. Knowing the mechanism of action of a nutraceutical(s), for instance, how essential nutrients work, is an important first criterion in verifying and promoting the role of nutraceuticals in better health. Determining the mechanism(s) of action of nutraceuticals is extremely important and an immense challenge.

Antioxidants can prevent free radical formation or can diffuse free radicals. The common nutraceutical antioxidants include the tetraterpenes (carotenoids and xanthophylls) in many vegetables and fruits and phenolics in a wider variety of plant foods. The tetraterpenes, specifically, act through their ability to quench singlet oxygen, thus preventing the initiation of free radicals. Lycopene is thought to be superior to β-carotene, and both have been reported to be superior to other carotenoids and vitamin E in their ability to quench singlet oxygen molecules.[165] Carotenoids and xanthophylls are lipid soluble and thought to function within cell membranes and lipoproteins, in contrast to phenolics, which are water soluble and perhaps most effective both outside and inside cells of the body. An exception is the theory that phenolics, especially those in red wine, are effective in preventing the oxidation of blood lipoprotein lipids. It is possible that these phenolics are protective to the lipoprotein while acting on its surface. While the tetraterpenes are thought to interact with singlet oxygen, the phenolics appear, based on their chemistry of multiple hydroxyl groups and availability to give up hydrogen atoms, to hydrogenate free radicals (hydroperoxyl, HOO•; hydroxyl, HO•; peroxyl, ROO• and; alkoxyl, RO•). Phenolics appear better at preventing the initiation and propagation of the oxidation reaction.

Antiinflammatory agents could act by preventing or regulating the overproduction of cytokines that could lead to autoimmune diseases.[166] The immune system elicits the production of cytokines to help fight off infections. It has been observed that omega-3 fatty acids act as antiinflammatory agents by reducing the level of these proinflammatory cytokines and thus may be helpful in the treatment of rheumatoid arthritis.[167]

The *antibacterial* actions of nutraceuticals should, logically, occur in the intestine, but the actions of various antibacterial compounds can be different. The sulfur compounds in garlic and onions appear effective against Gram-negative and Gram-positive bacteria.[168] The potential for these sulfur compounds to inhibit the growth and colonization of *Helicobacter pylori* could have great potential in reducing the incidence of stomach cancer[169] and gastric ulcers.[170] The sulfur compounds in the *Allium* family of plant foods could act through direct chemical reactions between their thiol groups and microorganisms.

Cranberries and blueberries and juices made from these fruits are recommended for the prevention of urinary tract infections (UTI). The active ingredients in these fruits are proanthocyanidins, which are phenolics in the subclass flavonoids. Urinary tract–pathogenic *E. coli* bacteria have the ability to adhere to mucosal surfaces because they have fimbriae, which are proteinaceous fibers, on their cell surfaces. The fimbriae of the pathogenic organism produce adhesins that aid attachment to uroepithelial cells by specific monosaccharide or oligosaccharide receptors. Proanthocyanidins are phenolic glycosides, and it has been proposed that the saccharide moieties on these phenolics attach to the organism, preventing its adherence to the mucosal lining.[171] The galactooligosaccharides (GOS), which are abundant in mother's milk, are believed to act in a similar manner, preventing the adhesion of pathogens to the infant's small intestine.[63]

Signaling can be described simply as communications within a cell, within an organ, or among cells and organs. Insulin and estrogen are classic examples of compounds involved in signaling in the body. A small amount of insulin produced by the pancreas and then circulated in the blood has the ability to signal almost all cells in the body to take up and utilize glucose. The phytoestrogens (phenolics) in soy and flax appear to compete with endogenous estrogen by binding to cells (e.g., breast tissue), thereby not allowing the excessive estrogenic activity (signaling) that is commonly associated with the increased incidence of breast cancer. Protein kinase exists in cell membranes, and cAMP produced by this enzyme is often referred to as the second message in the cell communication–signaling process. Sphingolipids change the structure of the cell membrane but also affect the activity of protein kinase and influence the potential signal activity of this enzyme and communication within the cell.

An explanation of the mechanism(s) that controls how nutrients and nutraceuticals affect *gene expression* may be among the most sought-after answers in the life sciences. Oncogenes are associated with the initiation of cancers in many parts of the body. It would be invaluable to know what nutraceuticals (or any chemical or drug) work most effectively as catalysts to effectively turn off these oncogenes. Likewise, enzymes that detoxify or destroy xenobiotics are important and include Phase I and Phase II enzymes in the liver. Although it is a noble and important idea to seek out nutraceuticals for cancer prevention, similar research is routinely accomplished in the search for new cancer drugs and drugs for other diseases.

Cell kinetics is the repetitive cycle (time) for cell reproduction that leads to cell growth and division into two daughter cells. There are four distinct events of the cell cycle: G_1 (GAP 1), S-Phase (DNA synthesis), G_2 (GAP 2), and M-Phase (mitosis). In phases G_1 and G_2, cells are simplistically described as being in a resting phase. Mitosis refers to the actual division of the cell. With relatively simple histological staining techniques, the two distinct nuclei can be observed just before the two new daughter cells are produced. It is S-Phase, when the cell's DNA is unraveled out from its double helix and in the process of replicating, that is thought to be most vulnerable for subsequent mutation through the addition or damaging action of carcinogens on the new DNA. There is a potential danger when cells reside in S-Phase for too long a time period. Their DNA is more susceptible to damage by carcinogens or by oxidation. The programmed death of a cell, called apoptosis, is associated with the cell cycle.[172] During the G_1 phase, when based on its environment, the cell is contemplating its future to replicate or destroy itself, it is highly advantageous for a cell to be killed off if the cell has the potential to malfunction or, as an example, mutate and become cancerous.

The production of butyrate, a short-chain fatty acid, resulting from the fermentation of dietary fiber, specifically some prebiotics and resistant starch, has been suggested to reduce the time intestinal mucosal cells are in S-Phase, thus reducing the period during which they are most susceptible to mutation.[173] Having intestinal mucosal cells replicate but reside in their S-Phase for shorter periods of time has been repeatedly suggested as one mechanism to explain the action of dietary fiber in preventing colorectal cancer.[174] The excessive production of reactive oxygen species (ROS) in the body is considered one of the important carcinogenic processes leading to oxidative damage of DNA with subsequent increased incidence of cancer.[175] Although the consumption of dietary omega-3 fatty acids, for example from fish oils, can result in higher levels of ROS,[176] these omega-3 fatty acids may help program damaged cells toward apoptosis and elimination.[177]

Hormonal action is a vital part of the communication system among bodily organs affecting possibly all physiological processes. This topic has been discussed under the heading of signaling. Some phenolic compounds have received attention as phytoestrogens. These include the isoflavonoids in soy, genistein and daidzein, and the lignans in flax.

Omega-3 fatty acids have been shown to change the ratio of eicosanoids produced from omega-6 fatty acids to the eicosanoids produced from omega-3 fatty acids, which can reduce platelet aggregation[178] and lower levels of proinflammatory cytokines.[179] Eicosanoids are called lipid hormones; they are cell signaling molecules and include prostaglandins, thromboxanes, leukotrienes, lipoxins, epoxides, hydroxyleicosatetraenoic acid = s (5-HETE and 12-HETE), and the diol form of HETE (diHETEs).

Energy for intestinal bacteria is necessary and is provided by fermentable carbohydrates such as different types of dietary fiber and soluble oligosaccharides of various sizes, having saccharide units of 3 to 10 and slightly larger. However, it is the fermentable carbohydrates, DP 3–10, which are receiving the most attention in promoting bacterial growth and which are defined as "prebiotics."[126] The concept behind increased consumption of prebiotics is to increase the growth of lactic acid–producing bacteria in the intestine, thus reducing the levels of pathogens that could be maintained at higher pH values. The importance of increased fermentation brought about by the increased consumption of fermentable dietary fiber, specifically prebiotics, is considered to be one of the most important health-associated benefits of dietary fiber consumption.

Immune stimulators or *immune enhancing nutrients* are thought to act first upon the primary and secondary immune system of the small intestine. Some of the more common immune enhancing essential nutrients and nutraceuticals are L-arginine, L-glutamine, nucleotides, and the long-chain fatty acids, which include eicosapentaenoic acid (EPA; C20:5 ω-3), and docosahexaenoic acid (DHA; C22:6 ω-3). Also included are the probiotics. Probiotics appear to show a tendency to stimulate the immune system by their presence and act upon the mucosal surface as they pass through the intestine. An interesting immune system stimulator is the chitin (poly *n*-acetyl glucosamine) polymer found in the product Quorn™.

Quorn is a food made from mycoprotein, which is produced by continuous fermentation of the microorganism *Fusarium graminearum* or *venenatumon* on carbohydrate substrates. The chitin in Quorn is like that in the walls of other fungi that can stimulate the intestinal immune system.[180] The polysaccharides of bacterial and fungal cell walls appear to be the active compounds stimulating the GALT.

Intestinal bulk pertains exclusively to the many insoluble sources of dietary fiber found in cereals, fruits, vegetables, and fractions thereof added to foods. These fractions or bulking sources of dietary fiber include isolated sources of cellulose, hemicellulose, and the lignin complex obtained by eating whole grains (ground), whole fruits, and vegetables. The most classic intestinal bulking agent is wheat bran, possibly followed by psyllium. One gram of wheat bran is reported to increase fecal mass by approximately 6 grams.[181] Psyllium is almost equally effective compared to wheat bran in the promotion of laxation and is commonly used as a powdered laxative, consumed after addition to water or orange juice (Metamucil®).[182] Insoluble sources of dietary fiber and a high-fiber diet are essential for laxation and both prevent and treat diverticulosis.[183]

Intestinal absorption can be modified to increase or decrease transport across the intestinal tract with a number of nutraceuticals. Animal protein will significantly enhance nonheme iron absorption in humans.[184] The complex of calcium carbonate–malic acid–citric acid (CCM) helps provide an approximate 10% increase in assimilation or bioavailability of the calcium in this complex compared to the calcium ions provided in supplements, such as calcium sulfate and calcium carbonate, and dairy products.[7] Plant sterols and stanols are effectively used to interfere with the absorption of dietary cholesterol in the gut, which helps to lower blood cholesterol levels.[185] There is speculation that enhanced nutrient absorption and fermentation of prebiotics, specifically FOS, and resistant starch, in the large intestine lead to increased absorption of calcium.[186,187] If this suggestion holds true, these fermentable supplements should have a tremendous impact for the prevention of osteoporosis.

Possibly the most notable *enzyme activity* system induced by a nutraceutical, specifically the glucosinolates, is the Phase I and Phase II detoxifying enzymes, or biotransformation enzymes, in the liver. The *Brassica* family of plants, which includes broccoli, Brussels sprouts, cabbage, cauliflower, kale, kohlrabi, and watercress, are rich sources of dietary glucosinolates. Although there are approximately 50 glucosinolates in these plant foods, two, glucobrassicin and glucoraphanin, are abundant in Brussels sprouts and have received the most attention. When crushed or broken, these glucosinolates are hydrolyzed by the enzyme thioglucoside glucohydrolase (myrosinase) to yield glucose, a sulfate ion, and isothiocyanates, thiocyanates, or indoles. Sulforaphane is obtained from glucoraphanin, and glucobrassicin can yield thiocyanic acid (CHNS), indole-3-carbinol (I3C), indole-3-acetic acid (IAA), indole-3-carboxylic acid (ICA), and 3,3'-diindolylmethane (DIM). Sulforaphane is considered to be a potent regulator of these detoxification enzymes, Phase I and Phase II.[188] Indole-3-carbinol may react with ascorbic acid to form ascorbigen, which is reported to have vitamin C activity, specifically, some type of antioxidant-like activity.[189] Although indole-3-carbinol is considered a major anticancer substance, and is commonly referred to as a member of the class of sulfur-containing compounds called glucosinolates, it does not contain sulfur.

The liver detoxifies harmful substances in two phases. The Phase I system, consisting of the P450 mixed-function oxidase enzymes, converts potentially toxic substances so that they can be rendered reactive to a series of Phase II enzymes.[190] However, Phase I enzymes can also convert procarcinogens to carcinogens. The Phase II enzymes use metabolites produced by the Phase I enzymes and add conjugates (e.g., glutathione, sulfate, glycone, acetate, cysteine, and glucuronic acid). These conjugates become biologically inactive and water soluble so they can be readily voided from the body via the urine. Two important Phase II enzymes are glutathione-S-transferase and quinine reductase. Glutathione (L-gammaglutamyl-L-cysteinylglycine) is a tripeptide consisting of the amino acids cysteine, glycine, and glutamic acid. Although glutathione is found in all plant and animal tissues, and is made in the body, it is also a popular antioxidant supplement. It is utilized by glutathione-S-transferase to donate hydrogen in the removal of ROS and serves as a conjugate to be added in the Phase II enzyme reaction. Sulforaphane is reported to be one of the most powerful dietary constituents capable of regulating the activity of these chemoprotective enzymes. It is possible that sulforaphane acts directly on the genes to induce the synthesis of these enzymes.

Prescription drugs to lower blood pressure are commonly referred to as ACE inhibitors. These drugs prevent the angiotensin-I converting enzyme (ACE) from converting angiotensin I into angiotensin II, which leads to high blood pressure.[191] Certain peptides in foods, specifically those found in whey protein hydrolyzates, have been found effective in lowering blood pressure and act as ACE inhibitors.[148–150]

The examples cited here to illustrate how nutraceuticals can function in the body to promote better health and prevent disease represent only a small fraction of those that occur. These examples are intended to help the reader expand this list.

CONCLUSION AND FUTURE CONSIDERATIONS

Foods have historically provided nourishment and enjoyment. Foods are cultural experiences. Possibly every culture has identified one or more foods and held this food in high regard for its perceived ability to promote health, enhance mental well being, or impart stamina. Many compounds in foods, other than essential nutrients, are now recognized as nutraceuticals — naturally occurring, health promoting chemicals. Nutraceuticals are the scientific revolution that has combined the disciplines of food science and nutrition in the pursuit of foods and their endogenous chemicals that can provide for better health and performance. Much remains to be learned about the many compounds in foods now described as nutraceuticals. Since each food is a functional food and can be a rich source of varied nutraceuticals, each food and its array of endogenous compounds deserve special attention. It will take time before an individual nutraceutical will receive the attention afforded essential nutrients. Nutritional scientists have spent the last century discovering and defining the actions of 41 essential nutrients. The value of nutraceuticals will be discovered in time. Although most foods exist as solids, almost all can be processed and delivered as beverages.

REFERENCES

1. Heasman, M. and Mellentin, J., *The Functional Foods Revolution, Healthy People, Healthy Profits?* Earthscan Publications Ltd., London, 2001.
2. Gordon, D.T., Seeking Your Opinions — Nutraceuticals, *Nutrition Notes*, June 2001, p. 4.
3. Gao, Y.T., McLaughlin, J.K., Blow, W.J., and Fraumeni, J.F., Reduced risk of esophageal cancer associated with green tea consumption, *J. Natl. Cancer Inst.*, 86, 855–858, 1994.
4. Fahey, J.W., Zhang, Y., and Talalay, P., Broccoli: an exceptionally rich source of inducers of enzymes that protect against chemical carcinogens, *Proc. Natl. Acad. Sci. USA*, 94, 10367–10372, 1997.
5. Humble, C., The evolving epidemiology of fiber and heart disease, in *Dietary Fiber in Health and Disease*, Kritchevsky, D. and Bonfield, C., Eds., Plenum Press, New York, 1997, pp. 15–26.
6. Anderson, J.W. and Bridges, S.R., Hypocholesterolemic effects of oat bran in humans, in *Oat Bran*, Wood, P.J., Ed., American Association of Cereal Chemists, Inc., St. Paul, 1993, pp. 139–157.
7. Andon, M.B., Peacock, M., Kanerve, R.L., and De Castro, J.A., Calcium absorption from apple and orange juice fortified with calcium citrate malate (CCM), *J. Am. Coll. Nutr.*, 15, 313–316, 1996.
8. Miettinen, T., Puska, P., Gylling, H., Vanhanen, H., and Vartiainen, E., Reduction of serum cholesterol with sitostanol-ester margarine in a mildly hypercholesterolaemic population, *N. Engl. J. Med.*, 333, 1308–1312, 1995.
9. Wise, J.W., Moring, R.J., Sanderson, R., and Bleum, K., Changes in plasma carotenoid, alpha-tocopherol, and lipid peroxide levels in response to supplementation with concentrated fruit and vegetable extracts: a pilot study, *Curr. Ther. Res.*, 57, 445–461, 1996.
10. Sloan, A.E., Top 10 trends to watch and work on. 3rd biannual report, *Food Technol.*, 55(4): 38–58, 2001.
11. Unrealized Prevention Opportunities — Reducing the Health and Economic Burden of Chronic Disease, United States Department of Health and Human Services, Centers for Disease Control and Prevention, 2000, pp. 1–45.
12. Bailey, L.B., Moyers, S., and Gregory, J.F., III, Folate, in *Present Knowledge in Nutrition*, 8th ed., Bowman, B. and Russell, R.M., Eds., ILSI Press, Washington, D.C., 2001, pp. 214–229.

13. Honein, M.A., Paulozzi, L.J., Mathews, T.J., Erickson, J.D., and Wong, L.Y., Impact of folic acid fortification of the U.S. food supply on the occurrence of neural tube defects, *JAMA,* 285, 2981–2986, 2001.
14. *Recommended Dietary Allowances,* 10th ed., Washington, D.C., National Academy Press, 1989.
15. Food and Nutrition Board, Institute of Medicine, *Dietary Reference Intakes for Thiamin, Riboflavin, Niacin, Vitamin B_6, Folate, Vitamin B_{12}, Pantothenic Acid, Biotin, and Choline; Dietary Reference Intakes for Calcium, Phosphorus, Magnesium, Vitamin D, and Fluoride; Dietary Reference Intakes for Vitamin C, Vitamin E, Selenium, and Carotenoids; Dietary Reference Intakes for Vitamin A, Vitamin K, Arsenic, Boron, Chromium, Copper, Iodine, Iron, Manganese, Molybdenum, Nickel, Silicon, Vanadium, and Zinc,* National Academy Press, Washington, D.C., 1999, 1999, 2000, and 2002.
16. Glanz, K., Basil, M., Mailbach, E., Goldberg, J., and Snyder, D., Why Americans eat what they do: taste, nutrition, cost, convenience and weight control concerns as influences on food consumption, *J. Am. Diet. Assoc.,* 98, 1118–1126, 1998.
17. Katz, F., Research priorities move toward healthy and safe, *Food Technol.,* 54(12), 42–46, 2000.
18. Institute of Medicine, National Academy of Sciences (Thomas, P.R., and Earl, R., Eds.), *Opportunities in Nutrition and Food Sciences,* National Academy Press, Washington, D.C., 1994, p. 109.
19. Guhr, G. and Lachance, P.A., Role of phytochemicals in chronic disease prevention, in *Nutraceuticals: Designer Foods III. Garlic, Soy and Licorice,* Lachance, P.A., Ed., Food & Nutrition Press, Inc., Trumbull, CT, 1997, pp. 311–364.
20. Federal Food, Drug and Cosmetic Act, June 25, 1938.
21. Dietary Supplement Health and Education Act (DSHEA), October 25, 1994.
22. Anderson, J.W., Deakins, D.A., and Bridges, S.R., Soluble fiber: hypocholesterolemic effects and proposed mechanisms, in *Dietary Fiber: Chemistry, Physiology, and Health Effects,* Kritchevsky, D., Bonfield, C., and Anderson, J.W., Eds., Plenum Press, New York, 1990, pp. 339–363.
23. Craig, S.A.S., Polydextrose: analysis and physiological benefits, in *Advanced Dietary Fibre Technology,* McCleary, B.V. and Prosky, L., Eds., Blackwell Science, London, 2001, pp. 503–508.
24. Clydesdale, F.M., Ed., First international conference on east–west perspectives on functional foods, *Nutr. Rev.,* 54(11), S1–S202, 1996.
25. Yalpani, M., Ed., *New Technologies for Healthy Foods and Nutraceuticals,* ATL Press, Inc., Shrewsbury, MA, 1997.
26. Mazza, G., *Functional Foods: Biochemical & Processing Aspects,* Technomic Publishing Co., Inc., Lancaster, PA, 1998.
27. Wildman, R.E.C., *Handbook of Nutraceuticals and Functional Foods,* CRC Press LLC, Boca Raton, FL, 2001.
28. Ohry, A. and Tsafrir, J., Is chicken soup an essential drug? *J. Can. Med. Assoc.,* 161, 1532–1533, 1999.

29. Shahm, N.P., Functional foods from probiotics and prebiotics, *Food Technol.*, 55(11), 46–53, 2001.
30. 21 CFR 101.30
31. 21 CFR 102.33
32. 21 CFR 101.3(b)(2)
33. Hasler, H.M., Functional foods: their role in disease prevention and health promotion, *Food Technol.*, 52(11), 63–70, 1998.
34. Health Canada, Health Canada Issues a Stop-Sale Order for All Products Containing Kava, advisory, Aug. 21, 2002.
35. Institute of Medicine, National Academy of Sciences, *Dietary Reference Intakes: A Risk Assessment Model for Establishing Upper Intake Levels for Nutrients*, National Academy Press, Washington, D.C., 1999.
36. The Kava-Kava in Food (England) Regulations, 2002.
37. Drewnowski, A. and Gomez-Carneros, C., Bitter taste, phytonutrients, and the consumer: a review, *Am. J. Clin. Nutr.*, 72, 1424–1435, 2000.
38. Doll, R., One for the heart, *Br. Med. J.*, 315, 1664–1668, 1997.
39. U.S. Departments of Agriculture and Health and Human Services, *Nutrition and Your Health: Dietary Guidelines for Americans*, 5th ed., U.S. Government Printing Office, Washington, D.C., 2000.
40. Poikolainen, K., Alcohol and mortality: a review, *J. Clin. Epidemiol.*, 488, 445–465, 1995.
41. Thun, M.J., Peto, R., Lopez, A.D., Monaco, J.H., Henley, J., Heath, C.W., and Doll, R., Alcohol consumption and mortality among middle-aged and elderly U.S. adults, *N. Engl. J. Med.*, 337, 1705–1714, 1997.
42. Frankel, E.N., Waterhouse, A.L., and Teissedre, P.L., Principal phenolic phytochemicals in selected California wines and their antioxidant activity in inhibiting oxidation of human low-density lipoproteins, *J. Agric. Food Chem.*, 43, 890–894, 1995.
43. Arichi, H., Kimura, Y., Okuda, H., Baba, K., Kozawa, M., and Arichi, S., Effects of stilbene components of the roots of *Polygonum cuspidatum Sieb. et Zucc.* on lipid metabolism, *Chem. Pharm. Bull.*, 30(5): 1766–1770, 1982.
44. Frankel, E.N., Kanner, J., and German, B., Inhibition of oxidation of human low-density lipoprotein by phenolic substances in red wine, *Lancet*, 341, 454–457, 1993.
45. Noble, A.C., Why do wines taste bitter and feel astringent? in *Chemistry of Wine Flavor*, Waterhouse, A.L. and Ebeler, S., Eds., American Chemical Society, Washington, D.C., 1998, pp. 156–165.
46. Gorinstein, S., Caspi, A., Zemser, M., and Trakhtenberg, S., Comparative contents of some phenolics in beer, red and white wines, *Nutr. Res.*, 20, 131–139, 2000.
47. Delage, E., Bohuon, G., Bason, A., and Drilleau, J.F., High-performance liquid chromatography of the phenolic compounds in the juice of some French cider apple varieties, *J. Chromatogr.*, 555, 125–136, 1991.

48. Anderson, J.W., Johnstone, B.M., and Cook-Newell, M.E., Meta-analysis of the effects of soy protein intake on serum lipids, *N. Engl. J. Med.*, 333, 276–282, 1995.
49. Lamartiniere, C.A., Protection against breast cancer with genistein, a component of soy, *Am. J. Clin. Nutr.*, 71, 1705S–1707S, 2000.
50. Lyle, B.J., Mares-Perlman, J.A., Klein, B.E., Klein, R., and Gregor, J.L., Antioxidant intake and risk of incident age-related macular cataracts in the Beaver Dam Eye Study, *Am. J. Epidemiol.*, 149, 801–808, 1999.
51. Seddon, J.M., Dietary carotenoids, vitamins A, C, and E, and advanced age-related macular degeneration. Eye disease case-control study group, *JAMA*, 272, 1413–1424, 1994.
52. Schmidt, M.K. and Labuza, T.P., Medical foods, *Food Technol.*, 46(4), 87–96, 1992.
53. Ganong, W.F., The general and cellular basis of medical physiology, in *Review of Medical Physiology*, 19th ed., Appleton & Lange, Stamford, CT, 1999, pp. 1–46.
54. Maughan, R.L. and Murray, R., Eds., *Sport Drinks: Basic Science and Practical Aspects*, CRC Press LLC, Boca Raton, FL, 2001.
55. Sloan, A.E., Top ten functional food trends, *Food Technol.*, 54(4), 33–62, 2001.
56. Yeum, K.-J. and Russell, R.M., Carotenoid bioavailability and bioconversion, *Annu. Rev. Nutr.*, 22, 484–504, 2002.
57. Hendrich, S. and Murphy, P.A., Isoflavones: source and metabolism, in *Handbook of Nutraceuticals and Functional Foods*, Wildman, R.E.C., Ed., CRC Press LLC, Boca Raton, FL, 2001, pp. 55–75.
58. Setchell, K.D.R., Brown, N.M., Desai, P., Zimmer-Nechemias, L., Wolfe, B.W., Brashear, W.T., Kirschner, A.S., Cassidy, A., and Heubi, J.E., Jr., Bioavailability of pure isoflavones in healthy humans and analysis of commercial soy isoflavone supplements, *J. Nutr.*, 131, 1362S–1375S, 2001.
59. Ross, J.A. and Kasum, C.M., Dietary flavonoids: bioavailability, metabolic effects, and safety, *Annu. Rev. Nutr.*, 22, 19–34, 2002.
60. Jelen, P. and Lutz, S., Functional milk and dairy products, in *Functional Foods: Biochemical & Processing Aspects*, Mazza, G., Ed., Technomic Publishing Co., Inc., Lancaster, PA, 1998, pp. 357–380.
61. Walzem, R.L., Dillard, C.J., and Geran, J.B., Whey components: millennia of evolution create functionalities for mammalian nutrition: what we know and what we may be overlooking, *Crit. Rev. Food Sci. Nutr.*, 42, 353–375, 2002.
62. Newburg, D.S., Human milk glycoconjugates that inhibit pathogens, *Curr. Med. Chem.*, 6, 117–127, 1999.
63. McVeagh, P. and Miller, J.B., Human milk oligosaccharides: only the breast, *J. Paediatr. Child Health*, 33, 281–286, 1997.
64. Farnworth, E.R., Probiotics and prebiotics, in *Handbook of Nutraceuticals and Functional Foods*, Wildman, R.E.C., Ed., CRC Press LLC, Boca Raton, FL, 2001, pp. 407–422.

65. Gorski, D., Kefir and Colostrum Beverage, *Dairy Foods*, Feb. 1998, p. 45.
66. Gorski, D., Probiotic Revolution, *Dairy Foods*, Jan. 1998, p. 41.
67. Heller, K.J., Probiotic bacteria in fermented foods: product characteristics and starter organisms, *Am. J. Clin. Nutr.*, 73, 374S–379S, 2001.
68. Schryver, T., The Soymilk Storm, *Nutraceuticals World*, Sept. 2000, p. 42.
69. Potter, S.M., Overview of possible mechanisms for the hypocholestrolemic effect of soy protein, *J. Nutr.* 125, 606S–611S, 1995.
70. Wu, A.H., Ziegler, R.G., Horn-Ross, P.L., Nomura, A.M.Y., West, D.W., Kolonel, L., Rosenthal, J.F., Hoover, R.N., and Pike, M.C., Soy intake and risk of breast cancer in Asian-Americans, *Am. J. Clin. Nutr.*, 68, 1437S–1443S, 1998.
71. Cassidy, A., Brigham, S., and Setchell, K.D.R., Biological effects of a diet of soy protein rich in isoflavones on the menstrual cycle of premenopausal women, *Am. J. Clin. Nutr.*, 60, 333–340, 1994.
72. Taylor, S.L. and Hefle, S.L., Food allergy, in *Present Knowledge in Nutrition*, 8th ed., Bowman, B. and Russell, R.M., Eds., ILSI Press, Washington, D.C., 2001, pp. 463–471.
73. Willett, W.C., Coffee consumption and coronary heart disease in women, *JAMA*, 275, 458–462, 1996.
74. Thompson, W.G., Coffee: brew or bane? *Am. J. Med. Sci.*, 308, 49–57, 1994.
75. Kleemola, P., Jousilahti, P., Pietinen, P., Vartiainen, E., and Tuomilehto, J., Coffee consumption and the risk of coronary heart disease and death, *Arch. Intern. Med.*, 160, 3393–3400, 2000.
76. Caffeine. A perspective on current concerns. Review from the National Institute of Nutrition in Canada, *Nutr. Today*, 1987, pp. 36–38.
77. Richelle, M., Tavazzi, I., and Offord, E., Comparison of the antioxidant activity of commonly consumed polyphenolic beverages (coffee, cocoa, and tea) prepared per cup serving, *J. Agric. Food Chem.*, 49, 3438–3442, 2001.
78. Clifford, M.N., Chlorogenic acids, in *Coffee, Vol. 1: Chemistry*, Clarke, R.J. and Macre, R., Eds., Elsevier, London, 1985, pp. 153–202.
79. Nardini, M., D'Aquino, M., Tomassi, G., Gentili, V., DeFelice, M., and Saccini, C., Effect of caffeic acid supplementation on the anti-oxidant defense system of the rat: an *in vivo* study, *Arch. Biochem. Biophys.*, 341, 157–160, 1997.
80. Nardini, M., Pisu, P., Gentili, V., Natella, F., DeFelice, M., Piccillela, E., and Saccini, C., Effect of caffeic acid on tert-butyl hydroperoxide–induced oxidative stress in u937, *Free Rad. Biol. Med.*, 25, 1098–1105, 1998.
81. Hertog, M.G.L., Feskens, E.J.M., and Hollman, P.C.H., Dietary antioxidant flavonoids and risk of coronary disease: the Zutphen elderly study, *Lancet*, 342, 1007–1011, 1993.
82. Katiyar, S.K., and Mukhtar, H., Tea in chemoprevention of cancer: epidemiologic and experimental studies, *Int. J. Oncol.*, 8, 221–238, 1996.
83. Kohlmier, L., Weterings, K.G.C., Steck, S., and Kok, F.J., Tea and cancer prevention: an evaluation of the epidemiologic literature, *Nutr. Cancer*, 27, 1–13, 1997.
84. Trevisanato, S.I. and Kim, Y.-I., Tea and health, *Nutr. Rev.*, 58, 1–10, 2000.

85. Mukhtar, H. and Ahma, N., Tea polyphenols: prevention of cancer and optimizing health, *Am. J. Clin. Nutr.,* 71, 1698S–1702S, 2000.
86. Leung, L.K., Su, Y., Chenc, R., Zhang, Z., Uang, Y., and Chen, Z.-U., Teaflavins in black tea and catechins in green tea are equally effective antioxidants, *J. Nutr.,* 131, 2248–2251, 2001.
87. Zhang, A., Chan, P.T., Luk, Y.S., Ho, W.K.K., and Chen, Z.Y., Inhibitory effect of jasmine green tea epicatechin isomers on LDL-oxidation, *J. Nutr. Biochem.,* 8, 334–340, 1997.
88. Cao, G., Sofic, E., and Prior, R.L., Antioxidant capacity of tea and common vegetables, *J. Agric. Food Chem.,* 44, 3426–3431, 1996.
89. Juneja, L.R.J., Chu, D.-C., Okubo, T., Nagato, Y., and Yokogoshi, H., L-Theanine: a unique amino acid of green tea and its relaxation effect in humans, *Trends Food Sci. Tech.,* 10, 199–204, 1999.
90. Hammerstone, J.F., Lazarus, S.A., Mitchell, A.E., Rucker, R., and Schmitz, H.H., Identification of procyanidins in cocoa (*Theobroma cacao*) and chocolate using high-performance liquid chromatography/mass spectrometry, *J. Agric. Food Chem.,* 47, 490–496, 1999.
91. Lazarus, S.A., Adamson, G.E., Hammerstone, J.F., and Schmitz, H.H., High-performance liquid chromatography/mass spectrometry analysis of proanthocyanidins in foods and beverages, *J. Agric. Food Chem.,* 47, 3693–3701, 1999.
92. Adamson, G.E., Lazarus, S.A., Mitchell, A.E., Prior, R.L., Cao, G., Jacobs, P.H., Kremers, B.G., Hammerstone, J.F., Rucker, R.B., Ritter, K.A., and Schmitz, H.H., HPLC method for the quantification of procyanidins in cocoa and chocolate samples and correlation to total antioxidant capacity, *J. Agric. Food Chem.,* 47, 4184–4188, 1999.
93. Kim, H. and Keeney, P.G., (–)Epicatechin content in fermented and unfermented cocoa beans, *J. Food Sci.,* 49, 1090–1092, 1984.
94. Ying, Y., Vinson, J.A., Etherton, T.D., Proch, J., Lazarus, S.A., and Kris-Ethertson, P.M., Effects of cocoa powder and dark chocolate on LDL oxidative susceptibility and prostaglandin concentrations in humans, *Am. J. Clin. Nutr.,* 74, 596–602, 2001.
95. Arts, I.C., Hollman, P.C., and Kromhout, D., Chocolate as a source of tea flavonoids, *Lancet,* 348, 834, 1999 (letter).
96. Achilli, G., Cellerino, G.P., and Gamache, P.H., Identification and determination of phenolic constituents in natural beverages and plant extracts by means of coulometric electrode array system, *J. Chromatogr.,* 632, 111–117, 1993.
97. Cheynbier, V., Souquet, J.-M., Le Roux, E., Guyot, S., and Rigaud, J., Size separation of condensed tannins by normal-phase high performance liquid chromatography, in *Methods in Enzymology, Vol 299, Oxidants and Antioxidants, Part A,* Parker, L., Ed., Academic Press, San Diego, 1999, pp. 351–357.
98. Nestel, P.J., Pomeroy, S., and Kay, S., Isoflavones from red clover improve systemic arterial compliance but not plasma lipids in menopausal women, *J. Clin. Endocrinol.,* 11, 43–47, 1999.

99. Duthie, G. and Crozier, A., Plant-derived phenolic antioxidants, *Curr. Opin. Lipidol.*, 11, 43–47, 2000.
100. Simpson, B.B. and Ogorzaly, M.C., *Economic Botany: Plants in Our World*, 3rd ed., McGraw-Hill, Inc., New York, 2001.
101. Holden, J.M., Eldridge, A.L., Beecher, G.R., Buzzard, M., Bhagwat, S., Davis, C.S., Douglass, L.W., Gebhardt, S., Haytowitz, D., and Sckakel, S., Carotenoid content of U.S. foods: an update of the database, *J. Food Comp. Anal.*, 12, 169–196, 1999.
102. Gould, M.N., Prevention and therapy of mammary cancer by monoterpenes, *J. Cell. Biochem.*, 22, 139–144, 1995.
103. Crowell, P.L. and Elson, C.E., Isoprenoids, health and disease, in *Handbook of Nutraceuticals and Functional Foods*, Wildman, R.E.C., Ed., CRC Press LLC, Boca Raton, FL, 2001, pp. 31–53.
104. Vigushin, D.M., Poon, G.K., Boddy, A., English, J., Halbert, G.W., Pagonis, C., Jarman, M., and Coombes, R.C., Phase I and pharmacokinetic study of D-limonene in patients with advanced cancer, *Cancer Chemother. Pharmacol.*, 42, 111–117, 1998.
105. Gutherie, N. and Carroll, K.K., Inhibition of mammary cancer by citrus flavonoids, in *Flavonoids in the Living System*, Manthey, J. and Buslig, B., Eds., Plenum Press, New York, 1998, pp. 44–55.
106. Rimm, E.B., Ascherio, A., Giovannucci, E., Spiegelman, D., Stampfer, M.J., and Willett, W.C., Vegetable, fruit, and cereal fiber intake and risk of coronary heart disease among men, *JAMA*, 275, 447–451, 1996.
107. Ness, A.R. and Powles, J.W., Fruit and vegetables and cardiovascular disease: a review, *Int. J. Epidemiol.*, 26, 1–13, 1997.
108. Giovannucci, E. and Clinton, S.K., Tomatoes, lycopene, and prostate cancer, *Proc. Soc. Exp. Biol. Med.*, 218, 129–139, 1998.
109. Giovannucci, E., Tomatoes, tomato-based products, and cancer: review of the epidemiologic literature, *J. Natl. Cancer Inst.*, 91, 317–331, 1999.
110. Food and Nutrition Board, Institute of Medicine, *Dietary Reference Intakes for Energy, Carbohydrate, Fiber, Fat, Fatty Acids, Cholesterol, Protein, and Amino Acids (Macronutrients)*, National Academy Press, Washington, D.C., 2002.
111. Todd, S., Woodward, M., Tunstall-Pedoe, H., and Bolton-Smith, C., Dietary antioxidant vitamins and fiber in the etiology of cardiovascular disease and all-cause mortality: results from the Scottish Heart Health Study, *Am. J. Epidemiol.*, 150, 1073–1080, 1999.
112. Fuchs, C.S., Giovannucci, E.L., Colditz, G.A., Hunter, D.J., Stampfer, M.J., Rosner, B., Speizer, F.E., and Willett, W.C., Dietary fiber and the risk of colorectal cancer and adenoma in women, *N. Engl. J. Med.*, 340, 169–176, 1999.
113. Schatzkin, A., Lanza, E., Corle, D., Lance, P., Iber, F., Caan, B., Shike, M., Weissfeld, J., Burt, R., Cooper, M.R., Kikendall, J.W., and Cahill, J., Lack of effect of a low-fat, high-fiber diet on the recurrence of colorectal adenomas, *N. Engl. J. Med.*, 342, 1149–1155, 2000.

114. Khachik, F., Beecher, G.R., and Smith, J., Jr., Lutein, lycopene, and oxidative metabolites in chemoprevention of cancer, *J. Cell Biochem.*, 22, 236–246, 1995.
115. Hirayama, O., Nakamura, K., Hamada, S., and Kobayasi, Y., Singlet oxygen quenching ability of naturally occurring carotenoids, *Lipids,* 29, 149–150, 1994.
116. Hennekens, C.H., Buring, J.E., Manson, J.E., Stampfer, M., Rosner, B., Cook, N.R., Belanger, C., LaMotte, F., Gaziano, J.M., Ridker, P.M., Willett, W., and Peto, R., Lack of effect of long-term supplementation with beta carotene on the incidence of malignant neoplasms and cardiovascular disease, *N. Engl. J. Med.,* 334, 1145–1149, 1996.
117. Omenn, G.S., Goodman, G.E., Thornquist, M.D., Balmesm, J., Cullen, M.R., Glass, A., Keogh, J.P., Meyskens, F.L., Valanis, B., Williams, J.H., Barnhart, S., and Hammar, S., Effects of a combination of beta carotene and vitamin A on lung cancer and cardiovascular disease, *N. Engl. J. Med.,* 334, 1150–1155, 1996.
118. Avorn, J., Monane, M., Gurwitz, J.H., Glynn, R.J., Choodnovskiy, I., and Lipsitz, L.A., Reduction of bacteriuria and pyuria after ingestion of cranberry juice, *JAMA,* 271, 751–754, 1994.
119. Nutrition and Athletic Performance — Position of the American Dietetic Association, Dietitians of Canada, and American College of Sports Medicine, *J. Am. Diet. Assoc.,* 100, 1543–1556, 2000.
120. Maughan, R.J., Fundamentals of sports nutrition: application to sports drinks, in *Sports Drinks: Basic Science and Practical Aspects,* Maughan, R.J. and Murray, R., Eds., CRC Press LLC, Boca Raton, FL, 2001, pp. 1–28.
121. Murray, R. and Stofan, J., Formulating carbohydrate–electrolyte drinks for optimal efficacy, in *Sports Drinks: Basic Science and Practical Aspects,* Maughan, R.J. and Murray, R., Eds., CRC Press LLC, Boca Raton, FL, 2001, pp. 197–223.
122. Coggan, A. and Coyle, E.F., Carbohydrate ingestion during prolonged exercise: effects on metabolism and performance, *Exerc. Sports Sci. Rev.,* 19, 1–40, 1991.
123. Below, P.R., Mora-Rodriguez, J., Gonzalez-Alonso, J., and Coyle, E.F., Fluid and carbohydrate ingestion independently improve performance during 1 h of intense exercise, *Med. Sci. Sports Exerc.,* 27, 200–210, 1994.
124. Nose, H., Mack, G.W., Shi, X., and Nadel, E.R., Role of osmolality and plasma volume during rehydration in humans, *J. Appl. Physiol.,* 65, 325–331, 1988.
125. Timothy, C. and Birdsall, N.D., Therapeutic applications of taurine, *Altern. Med. Rev.,* 3, 128–136, 1998.
126. Gibson, G.R., and Roberfroid, M.B., Dietary modulation of the human colonic microbiota: introducing the concept of prebiotics, *J. Nutr.,* 125, 1401–1412, 1995.
127. Gordon, D.T. and Cho, S., Eds., *Dietary Fiber, Prebiotics, Probiotics and Their Role in Intestinal Health,* Marcel Dekker, New York, (in press).

128. Gordon, D.T., Topp, K., Shi, Y.C., Zallie, J., and Jeffcoat, R., Resistant starch: physical and physiological properties, in *New Technologies for Healthy Foods & Nutraceuticals,* Yalpani, M., Ed., ATL Press, Shrewsbury, MA, 1997, pp. 157–178.
129. Carabin, I.G. and Flamm, W.G., Evaluation of safety of inulin and oligofructose as dietary fiber, *Reg., Toxicol. Pharmacol.,* 30, 268–282, 1999.
130. Brin, M. and Miller, O.N., The safety of oral xylitol, in *Sugars in Nutrition,* Sipple, H.L. and McNutt, K.W., Eds., Academic Press, New York, 1974, pp. 591–606.
131. Thomas, D.W., Edwards, J.B., and Edwards, R.G., Examination of xylitol, *N. Engl. J. Med.,* 283, 437–441, 1970.
132. Bertelsen, H., Jensen, B.B., and Buemann, B., D-Tagatose: a novel low-calorie bulk sweetener with prebiotic properties, *World Rev. Nutr. Diet.,* 85, 98–109, 1999.
133. Roberfroid, M.B., Van Loo, J.A., and Gibson, G.R., The bifidogenic nature of chicory inulin and its hydrolysis products, *J. Nutr.,* 128, 11–19, 1998.
134. Cashman, K., Prebiotics and calcium bioavailability, in *Probiotics and Prebiotics: Where Are We Going?* Tannock, G.W., Ed., Caister Academic Press, Norfolk, U.K., 2002, pp. 149–174.
135. Andon, M., *Super Calcium Miracle: The Calcium Citrate Malate Breakthrough,* Prima Publishing, Rocklin, CA, 1998.
136. Lanoue, L., Strong, P.L., and Keen, C.L., Adverse effects of low boron environment on the preimplementation development of mouse embryos *in vitro, J. Trace Elem. Exp. Med.,* 12, 235–250, 1999.
137. Chen, R.W. and Chuang, D.M., Long term lithium treatment suppresses p53 and Bax expression but increases Bcl-2 expression: a prominent role on neuroprotection against excitotoxicity, *J. Biol. Chem.,* 274, 6039–6042, 1999.
138. Offenbacher, E.G., Pi-Sunyer, F.X., and Stroecker, B.J., Chromium, in *Handbook of Nutritionally Essential Mineral Elements,* O'Dell, B.L. and Sunde, R.A., Eds., Marcel Dekker, Inc., New York, 1997, pp. 389–411.
139. Belury, M.A., Dietary conjugated linoleic acid in health: physiological effects and mechanism of action, *Annu. Rev. Nutr.,* 22, 505–531, 2002.
140. Chenoy, R., Hussain, S., Tayob, Y., O'Brien, P.M., Moss, M.Y., and Morse, P.F., Effect of oral gamolenic acid from evening primrose oil on menopausal flushing, *Br. Med. J.,* 308, 501–503, 1994.
141. Harris, W.S., n-3 fatty acids and serum lipoproteins: human studies, *Am. J. Clin. Nutr.,* 65, 1645S–1654S, 2000.
142. Connor, W.E., Importance of n-3 fatty acids in health and disease, *Am. J. Clin. Nutr.,* 71, 171S–175S, 2000.
143. Vesper, H.L., Schmelz, E.-M., Nikolova-Karakashian, M.N., Dillehay, D.L., Lynch, D.V., and Merrill, A.H., Jr., Sphingolipids in food and the emerging importance of sphingolipids to nutrition, *J. Nutr.,* 129, 1239–1250, 1999.
144. Clinton, S.K., Lycopene: chemistry, biology, and implications for human health and disease, *Nutr. Rev.,* 56(2 pt 1), 35–51, 1998.

145. Snodderly, D.M., Evidence for protection against age-related macular degeneration by carotenoids and antioxidant vitamins, *Am. J. Clin. Nutr.,* 62, 1448S–1461S, 1995.
146. Oakenfull, D. and Potter, J.D. 2001. Determination of the saponin content of foods, in *CRC Handbook of Dietary Fiber in Human Nutrition,* 3rd ed., Spiller, G.A., Ed., CRC Press LLC, Boca Raton, FL, 2001, pp. 127–130.
147. Gordon, D.T. and Godber, J.S., The enhancement of non-heme iron bioavailability by beef protein in the rat, *J. Nutr.,* 119, 446–452, 1989.
148. Nakamura, Y., Yamamoto, N., Sakai, K., Okubo, A., Yamazaki, S., and Takano, T., Purification and characterization of angiotensin I– converting enzyme inhibitors from sour milk, *J. Dairy Sci.,* 78, 777–783, 1995.
149. Nakamura, Y., Yamamoto, N., Sakai, K., and Takano, T., Antihypertensive effect of sour milk and peptides isolated from it that are inhibitors to angiotensin I –converting enzyme, *J. Dairy Sci.,* 78, 1253–1257, 1995.
150. Masuda, O., Nakamura, Y., and Takano, T., Antihypertensive peptides are present in aorta after oral administration of sour milk containing these peptides to spontaneously hypertensive rats, *J. Nutr.,* 126, 3063–3068, 1996.
151. Schloerg, P.R., Immune-enhancing diets: products, components, and their rationales, *J. Parenter. Enter. Nutr.,* 25(2), S3–S7, 2001.
152. Bravo, L., Polyphenols: chemistry, dietary sources, metabolism, and nutritional significance, *Nutr. Rev.,* 56, 317–333, 1998.
153. Kitts, D.D., Yuan, Y.V., Wijewickreme, A.N., and Thompson, L.U., Antioxidant activity of the flaxseed lignan secoisolariciresinol diglycoside and its mammalian lignan metabolites enterodiol and enterolactone, *Mol. Cell. Biochem.,* 202, 91–100, 1999.
154. Aman, P., Nilsson, M., and Anderson, R., Positive health effects of rye, *Cereal Foods World,* 42, 664–668, 1997.
155. Singleton, V.L., Naturally occurring food toxicants: phenolic substances of plant origin common in foods, *Adv. Food Res.,* 27, 149–242, 1981.
156. Guidelines for the Evaluation of Probiotics in Food. Report of a Joint FAO/WHO Working Group on Drafting Guidelines for the Evaluation of Probiotics in Food, London, Ontario, April 30 and May 1, 2002.
157. Teitelbaum, J.E. and Walker, W.A., Nutritional impact of pre- and probiotics as protective gastrointestinal organisms, *Annu. Rev. Nutr.,* 22, 107–138, 2002.
158. Sanders, M.E. and Klaenhammer, T.R., Invited review: the scientific basis of *Lactobacillus acidophilus* NCFM functionality as a probiotic, *J. Dairy Sci.,* 84, 319–331, 2001.
159. Vanderhoof, J.A., Young, R.J., Murray, N., and Kaufman, S.S., Treatment strategies for small bowel bacterial overgrowth in short bowel syndrome, *J. Pediatr. Gastroenterol. Nutr.,* 27, 155–160, 1998.
160. Klinman, D.M., Yi, A.K., Beaucage, S.L., Conover, J., and Kreig, A.M., CpG motifs present in bacterial DNA rapidly induce lymphocytes to secrete interleukin 6, interleukin 12, and interferon γ, *Proc. Natl. Acad. Sci. USA,* 93, 2879–2883, 1996.

161. Elmer, G.W., Probiotics: living drugs, *Am. J. Health-Syst. Pharm.*, 58, 1101–1109, 2001.
162. Tannock, G.W, Munro, K., Harmsen, H.J.M, Welling, G.W., Smart, J., and Gopal, P.K., Analysis of the fecal microflora of human subjects consuming a probiotic containing *Lactobacillus rhamnosus* DR20, *Appl. Environ. Microbiol.*, 66, 2578–2588, 2000.
163. Jeffery, E.H. and Jarrell, V., Cruciferous vegetables and cancer prevention, in *Handbook of Nutraceuticals and Functional Foods,* Wildman, R.E.C., Ed., CRC Press LLC, Boca Raton, FL, 2001, pp. 169–191.
164. Bowman, B.A. and Russell, R.M., Eds., *Present Knowledge in Nutrition,* 8th ed., ILSI Press, Washington, D.C., 2001.
165. Burton, G., Antioxidant actions of carotenoids, *J. Nutr.*, 119, 109–111, 1989.
166. Bendich, A., Carotenoids and the immune response, *J. Nutr.,* 119, 112–115, 1989.
167. Cooper, A.L., Gibbons, L., Horan, M.A., Little, R.A., and Rothwell, N.J., Effect of dietary fish oil supplementation on fever and cytokine production in human volunteers, *Clin. Nutr.,* 12, 312–328, 1993.
168. Yoshida, H., Katsuzaki, H., Ohta, R., Ishikawa, K., Fukuda, H., Fujin, T., and Suzuki, A., An organosulfur compound isolated from oil-macerated garlic extract, and its antimicrobial effects, *Biosci. Biotechnol. Biochem.,* 63, 588–590, 1999.
169. Cellini, L., DiCampli, E., Masulli, M., Di Bartolomeo, S., and Allocati, N., Inhibition of *Helicobacter pylori* by garlic extract (*Allium sativum*), *FEMS Immunol. Med. Microbiol.,* 13, 273–277, 1996.
170. El-Omar, E.M., Chow, C.H., and Rabkin, C.S., Gastric cancer and *H. pylori:* host genetics open the way, *Gastroenterology,* 121, 1002–1004, 2001.
171. Howell, A.B., Inhibition of the adherence of P-fimbriated *Escherichia coli* to uroepithelial cell surfaces by proanthocyanidin extracts from cranberries, *N. Engl. J. Med.,* 339, 1085–1086, 1998.
172. Refaeli, Y., Van Parijs, L., London, C.A., Tschopp, J., and Abbas, A.K., Biochemical mechanisms of IL-2–regulated Fas-mediated T cell apoptosis, *Immunity,* 8, 615–624, 1998.
173. van Munster, I.P. and Nagengast, F.M., Resistant starch decreases faecal secondary bile acids, cytotoxicity of faecal water and colonic mucosal proliferation, in *Intestinal Cell Proliferation with Emphasis on Dietary Manipulation,* Gee, J.M. and Nagengast, F.M., Eds., EURESTA Physiological Implication of the Consumption of Resistant Starch in Man — Contract No. AGRF/0027, 1992, pp. 35–36.
174. Gee, J.M. and Nagengast, F.M., Eds., *Intestinal Cell Proliferation with Emphasis on Dietary Manipulation,* EURESTA Physiological Implication of the Consumption of Resistant Starch in Man — Contract No. AGRF/0027, 1992.
175. Wiseman, H. and Halliwell, B., Damage to DNA by reactive oxygen and nitrogen species: role in inflammatory disease and progression to cancer, *Biochem. J.,* 313, 17–29, 1996.

176. Loft, S. and Poulsen, H.E., Cancer risk and oxidative DNA damage in man, *J. Mol. Med.*, 74, 297–312, 1996.
177. Mates, J.M. and Sanchez-Jimenez, F.M., Role of reactive oxygen species in apoptosis: implications for cancer therapy, *Int. J. Biochem. Cell Biol.*, 32, 157–170, 1999.
178. Knapp, H.R., Dietary fatty acids in human thrombosis and hemostasis, *Am. J. Clin. Nutr.*, 65, 1678S–1698S, 1997.
179. Endres, S. and von Schachy, C., n-3 Polyunsaturated fatty acids and human cytokine synthesis, *Curr. Opin. Lipidol.*, 7, 48–52, 1996.
180. Lim, B.O., Yamada, K., Nonaka, M., Kuramoto, Y., Hung, P., and Sugano, M., Dietary fibers modulate indices of intestinal immune function in rats, *J. Nutr.*, 127, 663–667, 1997.
181. Cummings, J.H., The effect of dietary fiber on fecal weight and composition, in *Dietary Fiber in Human Nutrition*, 3rd ed., Spiller, G.A., Ed., CRC Press LLC, Boca Raton, FL, 2001, pp. 183–252.
182. McRorie, J., Kesler, J., Bishop, L., Filloon, T., Allgood, G., Sutton, M., Hunt, T., Laurent, A., and Rudolph, C., Psyllium is superior to docusate sodium for treatment of chronic constipation, *Aliment. Pharmacol. Ther.*, 12, 491–497, 1998.
183. Brobribb, A.J.M., Treatment of symptomatic diverticular disease with a high-fibre diet, *Lancet*, i, 664–666, 1977.
184. Cook, D. and Monsen, E.R., Food iron absorption. I. Use of a semisynthetic diet to study absorption of food iron in man, *Am. J. Clin. Nutr.*, 28, 1289–1295, 1975.
185. Morohashi, T., Sano, T., Ohta, A., and Yamada, S., True calcium absorption in the intestine is enhanced by fructooligosaccharide feeding in rats, *J. Nutr.*, 128, 1815–1818, 1998.
186. Weststrate, J.A. and Meijer, G.W., Plant sterol-enriched margarines and reduction of plasma total- and LDL-cholesterol concentrations in normocholesterolaemic and mildly hypercholesterolaemic subjects, *Eur. J. Clin. Nutr.*, 52, 334–343, 1998.
187. Coudray, C., Bellanger, J., Dastiglia-Delavaud, C., Remesy, C., Vermorel, M., and Rayssignuier, Y., Effect of soluble or partly soluble dietary fiber supplementation on absorption and balance of calcium, magnesium, iron and zinc in healthy young men, *Eur. J. Clin. Nutr.*, 511, 375–380, 1997.
188. Zhang, Y., Talalay, P., Cho, C.-G., and Posner, G.H., A major inducer of anticarcinogenic protective enzymes from broccoli: isolation and elucidation of structure, *Proc. Natl. Acad. Sci. USA*, 89, 2399–2403, 1992.
189. Matano, K. and Kato, N., Studies on synthetic ascorbigin as a source of vitamin C for guinea pigs, *Acta Chem. Scand.*, 21, 2886–2891, 1967.
190. Prochaska, H.J., Santamaria, A.B., and Talalay, P., Rapid detection of inducers of enzymes that protect against carcinogens, *Proc. Natl. Acad. Sci. USA*, 89, 2394–2398, 1992.
191. Skeggs, L.T., Kahn, J.K., and Shumwaym, N.P., The preparation and function of the angiotensin-converting enzyme, *J. Exp. Med.*, 103, 295–299, 1956.

192. *The PDR of Herbal Medicines,* Medical Economics Corporation, Montvale, N.J., 1998.
193. Hoffman, D., *The Complete Illustrated Herbal,* Barnes & Noble Books, New York, 1999.
194. Health Products for Seniors: Anti-Aging Products Pose Potential for Physical and Economic Harm, GAO Report to Chairman, Special Committee on Aging, U.S. Senate, U.S. General Accounting Office, Washington, D.C., 2001 (GAO-01–1129).
195. Field, C.J., Use of T-cell function to determine the effect of physiologically active food components, *Am. J. Clin. Nutr.,* 71, 1720S–1725S, 2000.
196. Chang, Y.C., Nair, M.G., and Nitiss, J.L., Metabolites of daidzein and genistein and their biological activities, *J. Nat. Prod.,* 58, 1901–1905, 1995.
197. Arora, D.S. and Kaur, J., Antimicrobial activity of spices, *Int. J. Antimicrob. Agents,* 12, 257–262, 1999.
198. Akiyama, T., Ishida, J., Nakagawa, S., Ogawara, H., Watanabe, S., Itoh, N., Shibuya, M., and Fukami, Y., Genistein: a specific inhibitor of tyrosine-specific protein kinase, *J. Biol. Chem.,* 262, 5592–5595, 1987.
199. Bertram, J.S., Carotenoids and gene regulation, *Nutr. Rev.,* 57, 182–191, 1999.
200. Hooper, L.V., Won, M.H., Thelin, A., Hansson, L., Falk, P.G., and Gordon, J.I., Molecular analysis of commensal host–microbial relationships in the intestine, *Science,* 291, 881–884, 2001.
201. Alberts, D.S., Einspahr, J., Ritenbaugh, C., Aickin, M., Rees-McGee, S., Atwood, J., Emerson, S., Mason-Liddel, N., Bettinger, L., Patel, J., Bellapravalu, S., Ramanujam, P.S., Phelps, J., and Clark, L., The effect of wheat bran fiber and calcium supplementation on rectal mucosal proliferation rates in patients with resected adenomatous colorectal polyps, *Cancer Epidemiol. Biomarkers Prev.,* 6, 161–167, 1997.
202. Guoyao, W., Meininger, C.J., and Knabe, D.A., Arginine nutrition in development, health and disease, *Curr. Opin. Clin. Nutr. Metab. Care,* 3, 59–66, 2000.
203. McCracken, V.J. and Gaskins, H.R., Probiotics and the immune system, in *Probiotics: A Critical Review,* Tannock, G.W., Ed., Horizon Scientific Press, Norfolk, U.K., 1999, pp. 85–112.
204. Wood, P.J. and Beer, M.U., 1998. Functional oat products, in *Functional Foods: Biochemical & Processing Aspects,* Mazza, G., Ed., Technomic Publ. Co., Inc., Lancaster, PA, 1998, pp. 1–37.

4 Alternative Processing Technologies for the Control of Spoilage Bacteria in Fruit Juices and Beverages

Purnendu C. Vasavada

CONTENTS

Introduction 73
Control of Microbial Contamination 74
    Preventive Measures in the Orchard 74
    Washing 74
    Preservatives 78
    Pasteurization 81
Nonthermal Alternative Processing Technologies 81
    High Pressure Processing 82
    Pulsed Electric Field 85
    Ultraviolet Light 86
Irradiation 87
    Microwaves 87
Summary 88
Acknowledgment 89
References 90

INTRODUCTION

Fruit juices and fruit-based beverages are mildly acidic products, usually containing fermentable sugars, organic acids, vitamins, and trace elements, and are subject to contamination by and growth of a variety of spoilage organisms, notably yeasts and molds. Recent reports of outbreaks of illness caused by the consumption of fruits or fruit juices contaminated with pathogenic microorganisms such as *Salmonella, E. coli* O157:H7, and *Cryptosporidium*[1-4] have caused great concern. In the aftermath of these outbreaks, the U.S. Food and Drug Administration (FDA) has issued a guidance document to minimize microbial food safety hazards in fresh and minimally processed fruits and vegetables and mandated a Hazard Analysis and Critical Control Point (HACCP) program to achieve a 5-log reduction of pathogenic organisms.[5] The FDA also issued regulations dealing with the warning label on any unpasteurized juices that have not received a 5-log reduction process and recently, published the Juice HACCP final rule on January 19, 2001.[6,7]

CONTROL OF MICROBIAL CONTAMINATION

Microbial contamination of fruit can occur at all stages of growth, harvesting, storage, and processing. The surfaces of fresh fruits are often contaminated with yeasts and molds. The use of over-mature, damaged, or fallen fruit contaminated with manure from grazing animals has been implicated in *Salmonella* and *E. coli* O157:H7 outbreaks. Control of microbial contamination of fruit and fruit juice involves care at all stages of production, including preharvest practices of planting, growing of fruit, harvesting, postharvest handling, washing, and cooling and storage.

PREVENTIVE MEASURES IN THE ORCHARD

Contamination of fruits with feces of animals such as deer,[8] seagulls,[9] and cattle and other ruminants[10] in the orchard by direct or indirect contact should be prevented, and fertilizing orchards with manure should be avoided.[11] Using "drops" and damaged fruit increases the potential for microbiological contamination, including contamination with *E. coli,* and therefore should be avoided.[11-13] Another important source of *E. coli* O157:H7 infections is drinking water. Waterborne transmission of *E. coli* O157:H7 as a source of infection in domestic animals is a concern to human health as well. Wang and Doyle[14] reported that *E. coli* O157:H7 is a hardy pathogen that can survive for long periods of time in water, especially at cold temperatures. In an outbreak, *E. coli* O157:H7 was recovered from multiple water sources, including a borehole, a standpipe, and water stored in the home.[15] Precautions should be taken when using untreated water for washing purposes.

WASHING

Washing, mechanical scrubbing, and the use of chemical sanitizers may result in considerable reduction in surface contamination (see Table 4.1). Peroxyacetic acid (1280 ppm) was effective in accomplishing more than a

TABLE 4.1
Effects of Different Chemicals in Reducing Bacteria on the Surface of Fruits

Chemicals/ Disinfectants	Concentration	Type of Bacteria (Inoculum)	Sample	Log_{10} Reduction
		Pathogenic Bacteria		
Acetic acid	2–5%	*E. coli* O157:H7	Strawberry	1.6
	5%		Apple	3.1
Peroxyacetic acid (Tsunami 100)	80 ppm	*E. coli* O157:H7	Apple	2.6
	80 ppm [a]			3
	1280 ppm [b]			5.5
Tween 80	100–200 ppm	*E. coli* O157:H7	Strawberry	1.1–1.2
Sodium phosphate	2–5%	*E. coli* O157:H7	Strawberry	1.6–1.9
Hydrogen peroxide (H_2O_2)	1–3%	*E. coli* O157:H7	Strawberry	1.2–2.2
	3%	*E. coli* O157:H7	Tomato	4
	6%	*Salmonella chester*	Apple — on skin (cut)	3–4
			Apple — on stem and calyx (cut)	1–2
Chlorine dioxide (Oxine)	5 ppm	*E. coli* O157:H7	Apple	3
	80 ppm			4.5
Sodium hypochlorite (NaOCl)	100–200 ppm	*E. coli* O157:H7	Strawberry	1.3
	200 ppm		Apple	2.1
Calcium hypochlorite (CaOCl)	1.76%	*S. chester*	Apple (cut)	1–2
	36%	*S. chester*	Apple (cut)	1–2
Chlorine phosphate buffer (Agclor 310 [200 ppm]/Decco buffer 312)	200 ppm	*E. coli* O157:H7	Apple	3
	3200 ppm			4.5

(continued)

TABLE 4.1 (CONTINUED)
Effects of Different Chemicals in Reducing Bacteria on the Surface of Fruits

Chemicals/ Disinfectants	Concentration	Type of Bacteria (Inoculum)	Sample	$\log_{10}$ Reduction
Phosphoric acid	0.3%	*E. coli* O157:H7	Apple	2.9–2.3
Trisodium phosphate	2%	*S. chester*	Apple	1–2
Produce wash solution	A mixture of water, oleic acid, glycerol, ethanol, potassium hydroxide, sodium bicarbonate, citric acid, and distilled grapefruit oil	*S.* population (spp.) (*S. agona*, *S. enteritidis*, *S. gaminara*, *S. montevideo*, *S. typhimurium*)	Tomato	2–4
Combination treatment (acetic acid followed by hydrogen peroxide)	5% 3%	*E. coli* O157:H7	Apple	2.4–2.5
		Nonpathogenic Bacteria		
Hycrogen peroxide (H_2O_2) (50–60°C)	(5%)	*E. coli*	Apple	2.5
Hydrogen peroxide (H_2O_2) + acidic surfactants (50–60°C)				3–4

Sodium hypochlorite (NaOCl)	200 ppm	E. coli ATCC 25922	Apples	0.5
		E. coli ATCC 25922	Apples (half)	1.9
		E. coli ATCC 23716		1.4
		E. coli ATCC 11775		1.7
		Enterobacter aerogenes		2.0

[a] Recommended sanitizer concentration.
[b] 16 times the recommended concentration.

Adapted from references 9, 18, 23–25.

5-log reduction of *E. coli* O157:H7 on apple surface. A 4.5-log reduction of *E. coli* O157:H7 was obtained using chlorine phosphate buffer (3200 ppm) and chlorine dioxide (80 ppm). Hydrogen peroxide (5% H_2O_2) was less effective in reducing *E. coli* levels; however, addition of acidic surfactants (50–60°C) caused a 3- to 4-log reduction of *E. coli* on apple.

Failure to wash fruits properly before processing is among the main reasons for contamination in fruit juice. Washing and brushing fruit before the juicing step is common in juice processing. According to one industry survey, 98% of orchards surveyed washed apples before crushing, 18% used a detergent-based fruit wash, 37% used sanitizer after washing, and 64% employed brushing in conjunction with washing.[11] Winniczuk (1994)[17] showed that the maximum cleaning efficacy of most fruit wash systems produces a 90 to 99% reduction in the population of microorganisms on a citrus fruit surface under optimum pilot plant situations, whereas less-than-optimum conditions may result in only a 60% reduction of fruit surface microflora. However, washing trials using water showed only a 1- or 2-log reduction in many experimental research studies.[18,19]

Conventional washing practices using chlorine and brushing only may be partially effective in controlling microbial contamination.[9] The pathogens contaminating the fruit are not always located on the surface[20] and are not always distributed uniformly, thus limiting the effectiveness of surface treatments. Kenney et al. (2001)[19] suggested that cells may be sealed within naturally occurring cracks and waxy cuts in platelets. These cells may be protected from disinfection and subsequently released when apples are eaten or pressed for cider production. Also, the 5-log inactivation of pathogens on the surface may not necessarily result in requisite reduction of pathogens in juice.[21,22] For example, Pao and Davis (1999)[22] reported that an application of a 5-log inactivation treatment to oranges resulted in a 1.5- to 2.0-log reduction in juice. They also demonstrated that an overall 5-log inactivation of *E. coli* on the surface of oranges resulted in only a 3.5-log reduction in the juice. Treatment of fruit and vegetables with disinfectants is more effective in removing pathogenic microorganisms than washing with water alone but still not reliable enough to completely eliminate pathogenic bacteria.

Preservatives

Another approach for controlling contamination, especially in processed product, is the use of preservatives. In an industry survey, just 12% of producers reported using preservatives; among them, 60% used potassium sorbate and 40% used sodium benzoate. Potassium sorbate has little effect in reducing *E. coli* O157:H7 in cider.[26] Although sodium benzoate was more effective than potassium sorbate on *E. coli* O157:H7,[26] the bacteria survived in refrigerated cider containing 0.1% sodium benzoate for 21 days.[26] Similarly, citric and malic acids had no bactericidal effect.[27] Comes and Beelman (2002)[27] indicated that a 5-log reduction of *E. coli* O157:H7 in apple cider can be achieved using a preservation treatment involving the addition of fumaric acid (0.15%, w/v) and sodium benzoate (0.05%, w/v) to apple cider, followed by holding the cider at 25°C for 6 h before 24 h of refrigeration at 4°C. The final pH after the addition of fumaric acid and sodium benzoate was between 3.2 and 3.4. The authors suggested that this intervention process is cost effective and could easily be incorporated into HACCP systems that are currently mandated for processing of fruit and vegetable juices by the FDA.[27]

The use of preservatives may change heat resistance of *E. coli* O157:H7. For example, potassium sorbate and sodium benzoate reduce the heat resistance of *E. coli* O157:H7, but benzoate is about eight times more effective than sorbate.[28,29] Dock et al. (2000)[29] stated that addition of sodium benzoate (0.2%) increased the z-value from about 6 to 26°C. This increase may result in a longer 5-log reduction time (higher 5D-values) at higher temperatures (i.e., 70°C) in cider with benzoate as compared to cider without additives. This has profound implications because processors who add benzoate to cider before processing may obtain less than the 5-log reduction of *E. coli* O157:H7 that would have occurred without any benzoate addition. Induction of acid resistance can also have wide-ranging effects on the ability of bacteria to resist other stresses such as heating, antimicrobials, and exposure to ultraviolet light.[22,30] While preservatives may have some merit for extending product shelf life, they cannot be relied upon to eliminate pathogens from fruit juice or cider.

The FDA guideline for minimizing microbiological hazards emphasizes five major areas:

1. Water quality
2. Manure/bio-solids
3. Worker hygiene
4. Field, facility, and transport sanitation
5. Trace back

By considering the potential sources of contamination and implementing an effective combination of good agricultural and manufacturing practices (GAPs) related to apple juice/cider production, growers can minimize the risk of microbiological contamination.

Several effective alternative processing technologies have been developed for controlling microbial contamination, especially contamination with pathogenic microorganisms. These include pasteurization, high hydrostatic pressure (HHP) or ultra-high pressure (UHP), ultraviolet (UV), and pulsed

TABLE 4.2
Time–Temperature Conditions for Pasteurization of Apple Cider/Juices in the U.S.

Method	Process Conditions	Sample	Target Organism	Log_{10} Reduction
Heat pasteurization	71.1°C/>3 sec	Single-strength apple juice, orange juice, white grape juice	E. coli O157:H7 Salmonella spp. Listeria monocytogenes	≥5
	71.1°C/6 sec	Apple cider	E. coli O157:H7 (cocktail) Salmonella spp. (cocktail) L. monocytogenes (cocktail)	5
	71.1°C/160°F for 11 min or 76.7°C/170°F for 2 min	Apple cider produced from Red Delicious apples	E. coli O157:H7 (cocktail) Salmonella spp. (cocktail) L. monocytogenes (cocktail)	5
	Wisconsin recommendation: 68.1°C for 14 sec	Apple cider	E. coli O157:H7 (cocktail) Salmonella spp. (cocktail) L. monocytogenes (cocktail)	5

Adapted from Liao, C.H. and Sapers, G.M., *J. Food Prot.*, 63, 876–883, 2000; Mazzotta, A.S., *J. Food Prot.*, 64, 315–320, 2001.

electric field (PEF).[31,35,57,58] The design and implementation of HACCP and application of alternative processes provide the current strategy for microbial control and ensuring shelf life and safety of fruit juice and beverages.[5–7]

PASTEURIZATION[16,31,32]

Pasteurization of fruit and vegetable juice products has been applied for many years to reduce the microbial population and thus extend shelf life and to kill pathogenic bacteria to ensure safety (Table 4.2). Contemporary pasteurization processes are designed to inactivate 99.999% (5 log) of the organisms present in fruit juice. However, according to some surveys, thermal processing of apple juice or cider is not a popular option because of the perceived negative effects of pasteurization on the natural flavor and color of the juice products. In a recent survey, the majority (78%) of cider producers in Virginia indicated that they do not pasteurize their cider.[16] In another survey, 88% of producers in Wisconsin reported that they did not heat-pasteurize their apple juice or cider.[12] Also, mandatory pasteurization may be cost prohibitive for many smaller operations, because the costs increase sharply as production capacity and number of days per year of processing decrease.[33] However, a consumer survey in Wisconsin indicated that 70 panelists of 192 (36%) preferred buying pasteurized cider, 32 panelists (17%) preferred unpasteurized cider, and 79 panelists (41%) indicated no preference.[12] While there may be a preference or justification for using alternative nonthermal processing technologies to reduce microbial contamination of juice, many experts believe that the use of a kill step such as pasteurization rather than prevention of contamination is the best means of eliminating *E. coli* O157:H7 from apple cider.[32]

NONTHERMAL ALTERNATIVE PROCESSING TECHNOLOGIES

Since traditional thermal processes, though effective in inactivating bacteria, can affect the quality of the finished product, the scientific community has stepped up efforts to identify and review the kinetics and use of nonthermal alternative processes.[31] Most notably, as a part of the five-year contract between the Institute of Food Technologists (IFT) and the FDA, a scientific review of these alternative processing technologies has considered many pertinent questions, including:

- What might be used to produce food products free from any public health hazard, and what are the critical control points?
- Which organism(s) of public health concern is (are) the most resistant to the process(es)?

- How do factors such as growth phase and growth conditions of organism(s), processing substrate or food matrix, the pathogenic organisms associated with specific foods, processing conditions, storage conditions, and potential storage abuse affect the determination of the most resistant organism(s) of concern for each alternative processing technology?
- How do users determine the effectiveness of an alternative processing technology?

These are but a few of the significant issues being addressed. The IFT/FDA scientific review of the alternative processing technologies that might be used for both pasteurization and sterilization includes high pressure processing (HPP), pulsed electric field (PEF), pulsed x-ray or ultraviolet light (UV), ohmic heating, inductive heating, pulsed light, combined ultraviolet light and low-concentration hydrogen peroxide, ultrasound, filtration, and oscillating magnetic fields.[31] Some nonthermal alternative processing technologies are listed in Table 4.3.

HIGH PRESSURE PROCESSING

The use of HPP and/or ultra high pressure (UHP) as a food preservation technique is well documented. This type of nonthermal processing is currently used in various parts of the world in the manufacture of a number of products, including fruit juices, fruit purees, and jams.[44] HPP involves subjecting either packaged or unpackaged foods and beverages to pressures between 100 and 800 MPa within a cylindrical pressure vessel. The equipment used for a batch HPP system also includes two end closures with restraints such as yoke threads, a low pressure pump, an intensifier that uses liquid from the low pressure pump to generate high pressure process fluid for system compression, and system controls and instrumentation.[31] These batch system steps are rearranged for use to treat unpackaged liquid foods, such as fruit juices, semicontinuously. Recent studies suggest that this emerging alternative technology can offer food processors a viable nonthermal approach to ensuring food safety goals by inactivating bacteria. Several researchers have studied the efficacy of various HPP treatments in inactivating microorganisms, especially the pathogens *E. coli* O157:H7 and *Salmonella*, in fruit juices.[36,44-46] Some of these studies have shown that the lower the food's pH, the higher the number of microorganisms inactivated by HPP, as has been observed with the inactivation rates of *E. coli* O157:H7.[36] Similarly, spoilage organisms such as yeast in fruits can be effectively inactivated by using HPP due to their inherent low pH. Parish[45] targeted *Saccharomyces cerevisiae* in a nonpasteurized low-pH (3.7) orange juice with HPP, and a reported D-value of 76 seconds for ascospores treated at pressures between

TABLE 4.3
Nonthermal Processing Methods of Inhibiting Pathogenic Bacteria in Fruit Juices

Method	Process Conditions (Treatment)	Sample	Target Organism	Log_{10} Reduction	Reference
High pressure processing (HPP)	303–507 MPa (44,000–73,500 lb/in.^{2}) for 20 sec/min	Refrigerated juices		NS Extend shelf life up to 30 days	35
	551 MPa (80,000 lb/in^2) for 30 sec	Fresh juices	E. coli O157:H7 Salmonella	5	35
	615 MPa, 15°C, 2 min	Apple juice	E. coli O157:H7 (cocktail)	0.41	36
		Orange juice		2.16	
		Grapefruit juice		8.34	
		Carrot juice		6.40	
	615 MPa, 15°C, 2 min	Apple juice	Salmonella serovars	3.92–8.62	36
		Orange juice	Salmonella serovars	6.91–8.73	
		Grapefruit juice	Salmonella serovars	8.09–8.66	
		Carrot juice	Salmonella serovars	5.06–7.81	
Pulsed electric fields (PEF)	270 J/pulse, 1.2 V/µm, 20 pulses, square wave, <30°C	Apple juice	Saccharomyces cerevisiae	4.2	37
	260 J/pulse, 1.2 V/µm, 90 µs, 6 pulses exponential decay, 4–10°C	Apple juice	Saccharomyces cerevisiae	4	38
	558 J/pulse, 2.5 V/µm, 5 pulses, <25°C	Apple juice	Saccharomyces cerevisiae	3–4	38

(continued)

TABLE 4.3 (CONTINUED)
Nonthermal Processing Methods of Inhibiting Pathogenic Bacteria in Fruit Juices

Method	Process Conditions (Treatment)	Sample	Target Organism	$\log_{10}$ Reduction	Reference
	28 J/ml, 5.0 V/μm, 2.5 μs, 2 pulses, 22–29.6°C	Apple juice	Saccharomyces cerevisiae	6	37
	−2.5 V/μm, 2–20 μs, ± 150 pulses, exponential decay, <30°C	Apple juice		7	
	0.675 V/μm, 5 pulses	Orange juice	Saccharomyces cerevisiae	5	39
	<1 joule/ml Flow rate 20 gpm	Orange juice	E. coli O157:H7 Salmonella typhimurium Listeria monocytogenes	5–6 7 7	35, 40
	30 kV/cm, 143 μs, 25°C	Orange juice	E. coli O157:H7	5.0	35
	40 kV/cm, 143 μs, 25°C	Orange juice	Spoilage microorganisms	5.0	35
UV	Wavelength 253.7 nm mercury lamps	Apple, orange, carrot juices	E. coli O157:H7 Salmonella Listeria monocytogenes	5.0	41, 42
Irradiation	1.8 kGy	Apple juice	E. coli O157:H7	5.0	43

NS = not specified

350 and 500 MPa.[37] The D-values for the native flora of the orange juice ranged from 3 to 74 seconds. Yeasts and Gram-positive and Gram-negative organisms were found to survive 1 to 300 seconds of HPP treatment.[28] In addition, UHP extends the shelf life of refrigerated juice by up to 30 days if the pressures range from 44,000 to 73,500 lb/in.^{2} for 20 seconds to 1 minute. One significant problem with fresh orange juice is limited shelf life due to cloud loss caused by the activity of several pectin methylesterase (PME) enzymes. High-pressure processing can inactivate spoilage microflora and reduce PME activity.[35]

Pulsed Electric Field

PEF processing involves the application of high-voltage pulses for just a few microseconds to food placed or flowing between two electrodes. The process destroys both pathogens and spoilage organisms through breakdown or rupturing the cell membrane. Pores become permanent in most vegetative cells treated above 15,000 V/cm. PEF inactivates bacterial spores by reducing the dipicolinic acid needed for spore germination. The components of a PEF system include: [31,35]

- A high-voltage power supply
- An energy storage capacitor
- A treatment chamber or chambers
- A pump to conduct food though the treatment chamber(s)
- A cooling device
- Voltage, current, and temperature measurement devices
- A computer to control operations

Two commercially available systems have been used in pilot studies.[31] The most common use of pulse electric field processing has focused on food preservation and product quality aims, including extending the shelf life of orange juice, apple juice, bread, milk, and liquid eggs. In fact, shelf-life studies show that the process can extend refrigerated shelf life of fresh citrus juice to beyond 60 days.[31] In terms of inactivating spoilage microbes in fruit juices and fruit-based beverages, PEF has proved efficient in some research. In pilot experiments, researchers using PEF achieved a 5-log reduction of *E. coli* O157:H7 and its nonpathogenic surrogate *E. coli* 8739 in apple cider in 143 microseconds at a field strength of 30 kV/cm and average temperature of 25°C (near ambient). Spoilage organisms in orange juice were reduced by 5 logs at a peak field intensity of 40 kV/cm for 60 microseconds.[35] PEF technology also has been applied to process citrus juices in a slightly different, energy-efficient low-voltage electric pulse process in which electricity is directly pulsed into the juice. Less than 1 joule per ml is applied to a process flow rate of 20 gpm, resulting in a 7-log reduction of *Listeria monocytogenes* and *Salmonella typhimurium* and a 5-log reduction of *E. coli* O157:H7 in fresh orange juice.[40] Acidic liquid products such as fruit juices and pumpable particulate-containing liquids offer the best opportunity for commercialization of PEF technology. A commercial system combining PEF processing and aseptic packaging is being designed by a university/industry consortium.[35] This system will reportedly be capable of processing juice at flow rates of 2000 l/h at 35 kV/cm for 50 microseconds.

ULTRAVIOLET LIGHT

The use of UV light may be a promising, low-cost alternative to pasteurization for treating fruit juices, primarily apple juice and cider, to reduce microbial counts and inactivate pathogens such as *E. coli* O157:H7 and *Cryptosporidium parvum*.[16,47] UV light processing involves the use of mercury lamps, which generate 90% of their energy at a wavelength of 253.7 nanometers. Exposure of bacteria to UV results in cross-linking of the thymine dimers of the DNA in the organism, preventing repair of injury and reproduction. Recently, a California processor filed a petition with the FDA to allow the UV process in conjunction with HACCP to ensure a 5-log reduction of pathogens in its fresh, refrigerated juices.[35] The system consists of modules enclosing a one-inch diameter Teflon tube and has a process capacity of 420 gph at an exposure of UV at 253.7 nm for approximately one minute. A double pass through the module reportedly results in a 5-log reduction of *E. coli* O157:H7, *Salmonella,* and *Listeria monocytogenes* in four different juices — apple, orange, carrot, and mixed vegetable — without affecting flavor.[35]

Minor variations in the manufacture of the tubes can alter the fluid dynamics and bactericidal efficacy of UV radiation. Researchers at Rutgers University have validated individual quartz tubes for the CiderSure UV pasteurizer designed for UV treatment of apple cider to ensure that each tube meets the requirement of 100,000-fold reduction in the target pathogen, *E. coli* O157:H7, as proposed by the FDA.[47]

All tubes used in CiderSure units in cider mills demonstrated at least a 5-log reduction of a nonpathogenic surrogate for *E. coli* O157:H7 in each of three trials in the lab; tubes that failed to meet this criterion were not sold to cider producers.

The UV light may be less effective when interference caused by turbidity and background microorganisms is present. Also, pulp and other particulates may create a shadowing effect, shielding juice from the pasteurizing effect of the UV.[27,48] These limitations notwithstanding, the UV process can yield a 4-log reduction in pathogenic bacteria[35] and may be used in conjunction with HACCP and sanitation to achieve the 5-log reduction required in fresh, refrigerated fruit and vegetable juices. Turbulent flow and laminar flow UV processes are expected to be commercialized in the near future.

IRRADIATION

In contrast to the extensive studies on irradiation to control pathogens in meat and poultry products, very few studies of the value of ionizing irradiation for the elimination of foodborne pathogens on or in fruit juice, fruits, and vegetables have been conducted.[49] Ionizing irradiation has been used to eliminate *E. coli* O157:H7 from apple cider/juice. Fetter et al. (1969)[50] found that doses up to a maximum of 5 kGy had no effect on the flavor of commercial orange, tomato, apricot, peach, pear, and grape juices. Kiss and Farkas (1970)[51] observed marked increases in storage life of apple juice concentrates by irradiation. A dose of 13 kGy did not affect taste or aroma and ensured a storage life of at least 10 days at room temperature. The irradiation of fruits and vegetables is approved by FDA to a maximum dose of 1 kGy for disinfestations.[49]

In 1998, Buchanan et al.[43] found that the D-value for *E. coli* O157:H7 in apple juice at 2°C was dependent on the level of suspended solids and ranged from 0.26 to 0.35 kGy. The authors concluded that a dose of 1.8 kGy should be sufficient to achieve a 5D inactivation of *E. coli* O157:H7. *E. coli* is relatively sensitive to ionizing radiation[43] and can be controlled by low-dose treatment with a ^{137}Cs source. Buchanan et al.[30] suggested that there was substantial variability in radiation resistance among *E. coli* strains and that while pH differences between 4.0 and 5.5 can affect the radiation sensitivity of enterohemorrhagic *E. coli* (EHEC) strains, the overall impact of pH is relatively minor compared to the biological variability in the microorganisms' radiation resistance and the effects of prior growth conditions. The manipulation of pH would have limited impact in relation to the direct inactivation of EHEC in foods by irradiation. The ability of prior irradiation to increase the inactivation of EHEC during subsequent refrigerated storage is influenced by pH.

Growth of *E. coli* in an acidic environment also increases the microorganisms' resistance to UV light. Acid resistance could increase the radiation resistance of *E. coli* by enhanced repair of DNA. The use of combination treatments is expected to be more effective both in eliminating pathogens and in retaining quality attributes of the product.[49]

MICROWAVES

The use of microwaves to heat food for commercial pasteurization and sterilization in order to enhance microbial safety is discussed here. Microwave heating refers to the use of electromagnetic waves of certain frequencies to generate heat in a material. Typically, microwave food processing uses the two frequencies of 2450 and 915 MHz. The 2450 MHz frequency is used for home ovens, and both are used in industrial heating.

Microwave heating for pasteurization and sterilization is preferred to conventional heating primarily because it is rapid and therefore requires less time to come up to the desired process temperature. Other advantages of microwave heating systems are that they can be turned on or off instantly and that the product can be pasteurized after being packaged. Microwave processing systems also can be more energy efficient.[52]

The greater penetration depth and faster heating rates associated with microwave heating have been recognized as potential factors to improve the retention of thermolabile constituents in liquid foods such as milk and fruit juices.[53] Several studies have been performed on microwave pasteurization of milk.[54,55] Microwave pasteurization of fruits and fruit juices, e.g., citrus juices, involving enzyme inactivation and microbial destruction, however, has not been fully explored.[53]

The application of microwave energy for destruction of spoilage microorganisms in single-strength apple juice was explored.[53] Destruction kinetics of *S. cerevisiae* and *Lactobacillus plantarum* during continuous-flow microwave heating follows typical first-order reaction showing a linear destruction rate on a logarithmic plot of survivors versus time. Contributions of lethality during CUT and CDT were accommodated for when obtaining kinetic parameters. The destruction rate increased with an increase in temperature. Both microorganisms were similarly shown to be easily inactivated by both microwave and thermal treatments. Within the range of temperatures and sample sizes employed in this study, continuous microwave heating conditions destroyed the microorganisms an order of magnitude faster than did conventional batch heating conditions. This suggests the existence of some enhanced thermal effects associated with microwaves, resulting in a higher rate of microbial destruction as compared to conventional heating.

Other alternative techniques for the control of postharvest fungal spoilage of fruits have been studied. Ionizing irradiation has been examined, and it was found that doses of 3.5 kGy did not completely control postharvest decay of apple, quince, onion, and peach but did delay the growth of *Penicillium expansum, Monilia fructigena, Botrytis aclada,* and *Rhizopus stolonifer.*[56]

SUMMARY

The spoilage and microbial contamination of fruit juices and fruit-based drinks remain a concern for the industry. However, several approaches and processes are available to minimize the risk of contamination with pathogens

TABLE 4.4
Comparison of Nonthermal Juice Processes

Process	Temperature	Enzyme Inactivation	Equipment Costs	Packaging
Pulsed electric field	Ambient (slight increase due to process)	None	High	Aseptic or hygienic packaging
UV light	Ambient	None	Low	Aseptic or hygienic packaging
Minimal thermal process	70°C for 6 sec	Minimum	Low	Aseptic or hygienic packaging
Batch high pressure	Ambient plus compression heating	Selective inactivation	High	In-container processes
Continuous high pressure	Ambient plus compression heating	Selective inactivation	High	Aseptic or hygienic packaging

Source: Adapted from Sizer, C.E. and Balasubramaniam, Y.M., *Food Technol.*, 53(10), 64–67, 1999.

and ensure the safety of fruit juice and fruit beverages (Table 4.4).[35,57,58] These include washing and surface decontamination treatments, good orchard practices, and application of pasteurization processes.

With the emerging use and availability of nonthermal alternative processing technologies, such as HPP/UHP, PEF, and UV light, prospects for greater control look good. The successful application of these processes will depend on, among other things, the cost of equipment and effectiveness of the process.[41] While a few processes are at or near production scale, many are pilot scale and need further development. Also, most of these processes are new inventions and thus must be subjected to appropriate validation tests. Another complicating factor is the regulations dealing with labeling (e.g., designation as "fresh" for a juice "pasteurized" by a nonthermal process) and premarket approval. As juice HACCP gets underway, pinpointing the critical process hazards and identifying effective control measures will become more important than ever before.

ACKNOWLEDGMENT

This is a contribution from the College of Agriculture, Food and Environmental Sciences, University of Wisconsin River Falls and the Cooperative

Extension Service of the University of Wisconsin. The technical assistance and collaboration of Dr. Dilek Heperkan of Istanbul Technical University during her sabbatical visit at UW River Falls are gratefully acknowledged.

REFERENCES

1. U.S. Centers for Disease Control and Prevention (CDC), Outbreak of *Escherichia coli* O157:H7 infections associated with drinking unpasteurized commercial apple juice — British Columbia, California, Colorado, and Washington, *MMWR*, 45, 875, 1996.
2. U.S. Centers for Disease Control and Prevention (CDC), *Salmonella typhimurium* outbreak traced to a commercial apple cider — New Jersey, *MMWR*, 24, 87–88, 1975.
3. Millard, P.S., Gensheimer, K.F., Addiss, D.G., et al., An outbreak of cryptosporidiosis from fresh-pressed apple cider, *JAMA*, 272, 1592–1596, 1994.
4. U.S. Centers for Disease Control and Prevention (CDC), Outbreaks of *Escherichia coli* O157:H7 infections and cryptosporidiosis associated with drinking unpasteurized apple cider — Connecticut and New York, *MMWR*, 46, 4–8, 1998.
5. U.S. Food and Drug Administration (FDA), Hazard analysis and critical control point (HACCP) procedures for the safe and sanitary processing and importing of juice, proposed rule, *Federal Register*, 63, 20450–20486, 1998.
6. U.S. Food and Drug Administration (FDA), Hazard analysis and critical control point (HACCP) procedures for the safe and sanitary processing and importing of juice, final rule, *Federal Register*, 66, 6138–6202, 2001.
7. Anderson, S., Recent FDA juice HACCP regulations, *Food Saf.*, 7, 18–25, 2001.
8. Rice, D.H., Hancock, D.D., and Besser, T.E., Verotoxigenic *Escherichia coli* O157:H7 colonization of wild deer and range cattle, *Vet. Rec.*, 137, 524, 1995.
9. Sapers, G.M., Miller, R.L., and Mattrazzo, A.M., Effectiveness of sanitizing agents in inactivating *Escherichia coli* I in golden delicious apples, *J. Food Sci.*, 64, 734–737, 1999.
10. Borcyzk, A.A., Lior, H., and Duncan, L.M.C., Bovine reservoir for verotoxin producing *Escherichia coli* O157:H7, *Lancet*, i, 98, 1987.
11. Wright, J.R., Sumner, S.S., Hackney, C.R., Pierson, M.D., and Zoecklein, B.W., A survey of Virginia apple cider producers' practices, *Dairy Food Environ. Sanit.*, 20, 190–195, 2000.
12. Uljas, H.E. and Ingham, S.C., Survey of apple growing, harvesting, and cider manufacturing practices in Wisconsin: implications for safety, *J. Food Saf.*, 20, 85–100, 2000.
13. Riordan, D.C.R., Sapers, G.M., Hankinson, T.R., Magee, M., Mattrazzo, A.M., and Annous, B.A., A study of U.S. orchards to identify potential sources of *Escherichia coli* O157:H7, *J. Food Prot.*, 64, 1320–1327, 2001.

14. Wang, G. and Doyle, M.P., Survival of enterohemorrhagic *E. coli* O157:H7 in water, *J. Food Prot.*, 61, 662–667, 1998.
15. Effler, P., Isaacson, M., Arntzen, I., Heenan, R., Canter, P., Barrett, T., Lee, L., Mambo, C., Levine, W., Zaidi, A., and Griffin, P.M., *Emerg. Infect. Dis.*, 7, 812–819, 2001.
16. Wright, J.R., Sumner, S.S., Hackney, C.R., Pierson, M.D., and Zoecklein, B.W., Efficacy of ultraviolet light for reducing *Escherichia coli* O157:H7 in unpasteurized apple cider, *J. Food Prot.*, 63, 563–567, 2000.
17. Winniczuk, P., Effects of Sanitizing Compounds on the Microflora of Orange Fruit Surfaces and Orange Juice, thesis, University of Florida, Gainesville, 1994.
18. Wisniewsky, M.A., Glatz, B.A., Gleason, M.L., and Reitmeier, C.A., Reduction of *Escherichia coli* O157:H7 counts on whole fresh apples by treatment with sanitizers, *J. Food Prot.*, 63, 703–708, 2000.
19. Kenney, S.J., Burnett, S.L., and Beuchat, L.R., Location of *Escherichia coli* O157:H7 on and in apples as affected by bruising, washing, and rubbing, *J. Food Prot.*, 64, 1328–1333, 2001.
20. Buchanan, R.L. and Edelson, S.G., pH-dependent stationary-phase acid resistance response of enterohemorrhagic *Escherichia coli* in the presence of various acidulants, *J. Food Prot.*, 62, 211–218, 1999.
21. Pao, S. and Davis, C.L., Enhancing microbiological safety of fresh orange juice by fruit immersion in hot water and chemical sanitizers, *J. Food Prot.*, 62, 756–760, 1998.
22. Pao, S. and Davis, C.L., Maximizing microbiological quality of fresh orange juice by processing sanitation and fruit surface treatments, *Dairy Food Environ. Sanit.*, 21, 287–291, 1999.
23. Wright, J.R., Sumner, S.S., Hackney, C.R., Pierson, M.D., and Zoecklein, B.W., Reduction of *Escherichia coli* O157:H7 on apples using wash and chemical sanitizer treatments, *Dairy Food Environ. Sanit.*, 20, 120–126, 2000.
24. Harris, L.J., Beuchat, L.R., Kajs, T.M., Ward, T.E., and Taylor, C.H., Efficacy and reproducibility of a produce wash in killing *Salmonella* on the surface of tomatoes assessed with a proposed standard method for produce sanitizers, *J. Food Prot.*, 64, 1477–1482, 2001.
25. Liao, C.H. and Sapers, G.M., Attachment and growth of *Salmonella cheter* on apple fruits and *in vivo* response of attached bacteria to sanitizer treatments, *J. Food Prot.*, 63, 876–883, 2000.
26. Miller, L.G. and Kaspar, C.W., *Escherichia coli* O157:H7 acid tolerance and survival in apple cider, *J. Food Prot.*, 57, 460–464, 1994.
27. Comes, J.E. and Beelman, R.B., Addition of fumaric acid and sodium benzoate as an alternative method to achieve a 5-log reduction of *E. coli* O157:H7 populations in apple cider, *J. Food Prot.*, 65, 476–483, 2002.
28. Splittstoesser, D.F., McLellan, M.R., and Churney, J.J., 1996. Heat resistance of *Escherichia coli* O157:H7 in apple juice, cited in Dock, L.L. et al., *J. Food Prot.*, 59, 226–229, 2000.

29. Dock, L.L., Floros, J.D., and Linton, R.H., Heat inactivation of *Escherichia coli* O157:H7 in apple cider containing malic acid, sodium benzoate and potassium sorbate, *J. Food Prot.*, 63, 1026–1031, 2000.
30. Buchanan, R.L., Edelson, S.G., and Boyd, G., Effects of pH and acid resistance on the radiation resistance of enterohemorrhagic *Escherichia coli, J. Food Prot.*, 62, 219–228, 1999.
31. Institute of Food Technologists, Kinetics of microbial inactivation for alternative food processing technologies, *J. Food Sci.* (suppl.), 65, 1–108, 2000.
32. Mak, P.P., Ingham, B.H., and Ingham, S.C., Validation of apple cider pasteurization treatments against *Escherichia coli* O157:H7, *Salmonella*, and *Listeria monocytogenes, J. Food Prot.*, 64,1679–1689, 2001.
33. Kozempel, M., McAloon, A., and Yee, W., The cost of pasteurizing apple cider, *Food Technol.*, 52(1), 50–52, 1998.
34. Mazzotta, A.S., Thermal inactivation of stationary-phase and acid-adapted *Escherichia coli* O157:H7, *Salmonella*, and *Listeria monocytogenes* in fruit juices, *J. Food Prot.*, 64, 315–320, 2001.
35. Morris, E., FDA regs spur non-thermal R&D, *Food Engineering*, July/August, 61–68, 2000.
36. Teo, A., Ravishankar, S., and Sizer, C.E., Effect of low-temperature, high-pressure treatment on the survival of *Escherichia coli* O157:H7 and *Salmonella* in unpasteurized fruit juices, *J. Food Prot.*, 64, 1122–1127, 2001.
37. Qin, B.L., Pothakamury, U.R., Vega, H., Martin, O., Barbosa-Canovas, G.V., and Swanson, B.G., Food pasteurization using high-intensity pulsed electric fields, *Food Technol.*, 55–60, 1995.
38. Zhang, Q.H., Gonzales, A.M., Barbosa-Canovas, G.V., and Swanson, B.G., Inactivation of *E. coli* and *S. cerevisiae* by pulsed electric fields under controlled temperature conditions, *Trans. ASAE*, 37, 581–587, 1994.
39. Grahl, T. and Markl, H., Killing of microorganisms by pulsed electric fields, *Appl. Microbiol. Biotechnol.*, 45, 148–157, 1996.
40. Pulse power disinfects fresh juices, extends shelf life, *Food Engineering*, October 1998, pp. 47–50.
41. UV light provides alternative to heat pasteurization of juices, *Food Technol.*, 53, 144, 1999.
42. UV light process extends shelf life, *Food Engineering*, Nov. 1999, pp. 14–15.
43. Buchanan, R.L., Edelson, S.G., Snipes, K., and Boyd, G., Inactivation of *Escherichia coli* O157:H7 in apple juice by irradiation, *Appl. Environ. Microbiol.*, 64(11): 4533–4535, 1998.
44. Linton, M., et al., Inactivation of *Escherichia coli* O157:H7 in orange juice using a combination of high pressure and mild heat, *J. Food Prot.*, 62, 277–279, 1999.
45. Parish, M.E., High pressure inactivation of *Saccharomyces cerevisiae*, endogenous microflora and pectinmethylesterase in orange juice, *J. Food Prot.*, 18, 57–65, 1998.
46. Garcia-Graells, C., Hauben, K.J.A., and Michiels, C.W., High-pressure inactivation and sublethal injury of pressure-resistant *Escherichia coli* mutants in fruit juices, *Appl. Environ. Microbiol.*, 64, 1566–1568, 1998.

47. Duffy, S., Churey, J., Worobo, R.W., and Schaffner, D.W., Analysis and modeling of the variability associated with UV inactivation of *Escherichia coli* in apple cider, *J. Food Prot.*, 63, 1587–1590, 2000.
48. Milligen, D.V., Sanitation 101, *Food Engineering*, January 2001, pp. 55–59.
49. Thayer, D.W. and Rajkowski, K.T., Developments in irradiation of fresh fruits and vegetables, *Food Technol.*, 53(11), 62–65, 1999.
50. Fetter, F., Stehlik, S., Kovacs, J., and Weiss, S., 1969, cited in Thayer, D.W. and Rajkowski, K.T., Developments in irradiation of fresh fruits and vegetables. *Food Technol.*, 53(11), 62–65, 1999.
51. Kiss, I. and Farkas, J. 1970, cited in Thayer, D.W. and Rajkowski, K.T., Developments in irradiation of fresh fruits and vegetables, *Food Technol.*, 53(11), 62–65, 1999.
52. Datta, A.K. and Davidson, P.M., Microwave and radio frequency processing, in Kinetics of microbial inactivation for alternative food processing technologies, Special supplement, *J. Food Sci.*, 65, 32–41, 2001.
53. Tajchakavit, S., Ramaswamy, H.S., and Fustier, P., Enhanced destruction of spoilage microorganisms in apple juice during continuous flow microwave heating, *Food Res. Int.*, 31, 713–722, 1998.
54. Knutson, K.M., Marth, E.H., and Wagner, M.K., Use of microwave ovens to pasteurize milk, *J. Food Prot.*, 51, 715–719, 1988.
55. Kudra, T., Van De Voort, F.R., Raghavan, G.S.V., and Ramaswamy, H.S., Heating characteristics of milk constituents in a microwave pasteurization system, *J. Food Sci.*, 56, 931–934, 937, 1991.
56. Tiryaki, O., Aydn, G., and Gürer, M., Post-harvest disease control of apple, quince, onion and peach with radiation treatment, *J. Turkish Phytopathol.*, 23, 143–152, 1994.
57. Vasavada, P.C. and Heperkan, D., Non-thermal alternative processing technologies for the control of spoilage bacteria in fruit juices and fruit-based drinks, *Food Safety Magazine*, 8(1): 8, 10, 13, 46–47, 2002.
58. Sizer, C.E. and Balasubramaniam, Y.M., New intervention processes for minimally processed juices, *Food Technol.*, 53(10), 64–67, 1999.

5 Microbiology of Fruit Juice and Beverages

Purnendu C. Vasavada

CONTENTS

Introduction 95
Microbial Spoilage of Fruit and Fruit Juice and Beverages......................98
    Bacteria 98
    Yeasts and Molds..101
    Indicator Bacteria and Pathogenic Organisms 102
    Protozoa 103
    Pathogenic Yeasts...103
    Viruses...103
    Mycotoxins 105
Emerging Pathogens and Outbreaks of Illness 105
    Early Outbreaks 105
    Outbreaks in the 1990s..106
    Illness from Other Potential Food Safety Hazards 109
Ensuring Safety of Juice: Strategy and Control..110
    GMP and Best Practices for Juice Processors 113
    Model HACCP...116
Summary...118
Acknowledgment 118
References..118

INTRODUCTION

Fruit juices and fruit-based beverages are popular products appealing to a broad demographic group, particularly children and young adults, and represent an important segment of the domestic and international market. Juices are the aqueous liquids expressed or otherwise extracted usually from one or more fruits or vegetables, purees of the edible portion of one or more fruits or vegetables, or any concentrates of such liquids or purees. Fruit juice may be an ingredient in beverages. A wide variety of juice and beverage

TABLE 5.1
New Beverage Introductions

	2001	2000	1999
Hot beverages	479	470	575
RTD juices/juice drinks	265	217	307
Concentrates/mixes	122	154	121
Energy/sports drinks	104	86	107
Carbonated soft drinks	82	68	94
RTD iced tea/coffee	43	102	80
Beer/cider	25	93	65
Water	50	65	65
Flavored alcoholic drinks	29	16	12
Total	1199	1271	1426

Source: Adapted from Enright, A., *Prepared Foods*, 170(4), 41–42, 2001; Roberts, W. and Dornblaser, L., *Prepared Foods,* 171(4), 19–29. 2002.

products including juice (100% juice), juice blends (combinations of several juices), juice drinks (not 100% juice), and flavored beverages in a variety of types, including fresh, refrigerated, shelf stable, frozen concentrate, nonfrozen concentrate, sports drinks, energy drinks, etc., are currently available on the U.S. market (see Table 5.1).[1-5] Consumption of fruit juice and beverages in the U.S. has increased steadily during the past two decades. In 1998, U.S. consumers drank an average of about 9 gallons of juice and about 6 gallons of fruit beverages per year.[2] While consumption of milk, coffee, wine, beer, and spirits has generally declined in recent years, consumption of juices, beverages, soft drinks, sports drinks, and bottled water has increased.[6] In 1999, consumers drank an average of 15.5 gallons of bottled water, 55.9 gallons of soft drinks, and 2.3 gallons of sports drinks.[3] Per capita consumption trends are summarized in Figure 5.1.[6,7]

Fruit juices and beverages are important commodities in the global market, providing ample opportunity for innovative, value added products to meet consumer demand for convenience, nutrition, and health. Beverages constituted a significant proportion (33–73%) of various health promoting new products or product lines introduced in the U.S. in 2000 (see Table 5.2).[4] According to a recent industry report, the U.S. functional beverage market generated revenues of $4.7 billion in 2000 and is expected to exceed $12 billion by 2007.[4]

Fruit juices and beverages contain water, sugars, organic acids, vitamins, and trace elements and provide an ideal environment for spoilage by microorganisms, particularly yeasts, molds, and aciduric organisms. Pathogenic bacteria are usu-

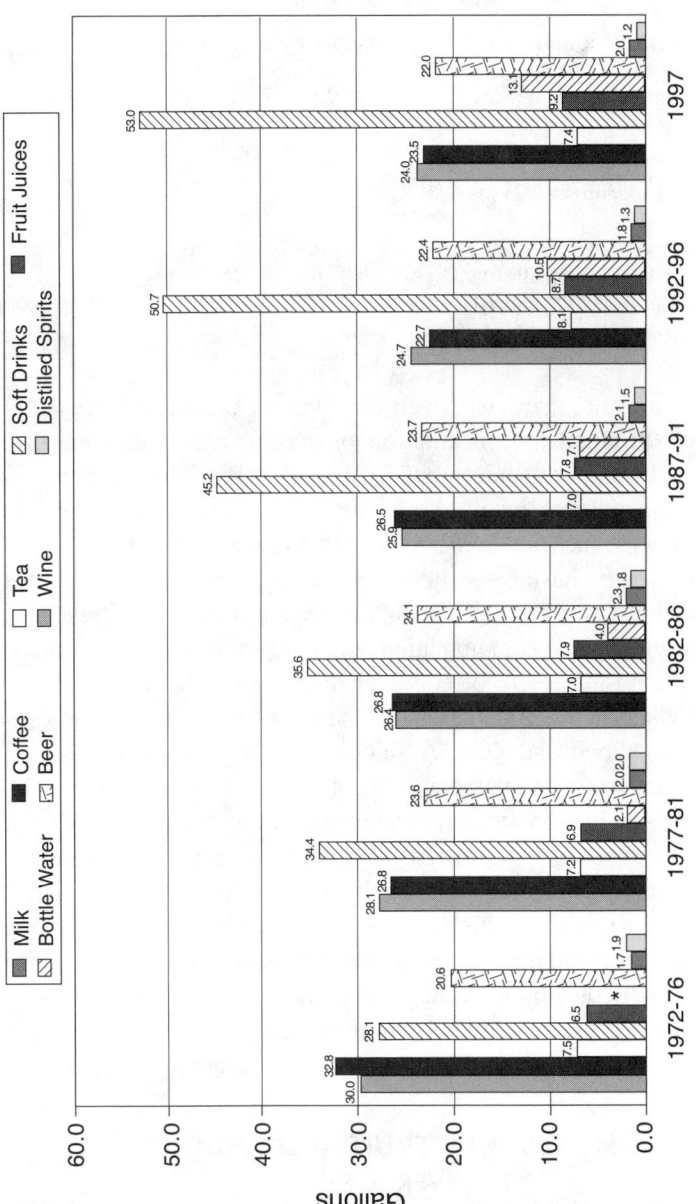

FIGURE 5.1 U.S. per capita beverage consumption: 1972–1997. * = No bottled water data collected during this period. (Adapted from Rowles, K., Processed apple products and marketing analysis: apple juice and cider, SP 2002–01, Cornell University, Ithaca, NY, 2001 and Putnam, J.J. and Allshouse, J.E., Food Consumption, Prices and Expenditure: 1970–1997, Food and Rural Economics Division, USDA Economic Research Service, Statistical Bulletin No. 965, April 1999.)

TABLE 5.2
New Products Introduced in the U.S. in 2000 Claiming Health Benefits

Health Condition	Beverages (%)
Immune (62)	73
Heart health (59)	56
Diabetes (24)	33
Osteoporosis (23)	70
Cancer (4)	50

Source: Adapted from O'Donnell, C.D., *Prepared Foods,* 170(4), 50–51, 2001.

ally not a problem in fruit juices and beverages. However, several outbreaks of foodborne illness attributed to consumption of commercial, nonpasteurized (unpasteurized or "fresh") fruit juices and beverages have occurred in recent years.[7–10] At least one outbreak involved a fatality.[11] These outbreaks, attributed to emerging pathogens such as *Escherichia coli* O157:H7, *Salmonella,* and *Cryptosporidium parvum,* have caused concern among the consuming public. While 98% of the juice sold in the U.S. is pasteurized or otherwise treated to control the risk of pathogenic contamination, some 40 million gallons of juice are not pasteurized, posing a risk of contamination with pathogenic bacteria. Unpasteurized products cause 6000 cases of illness per year, according to estimates from the U.S. Food and Drug Administration (FDA). To address the problem of pathogenic contamination in fruit juice and beverages, the FDA issued regulations, as an interim measure, requiring warning labels on juices that are "fresh" or have not been processed to destroy pathogens that may be present.[13] The FDA also required implementation of the Hazard Analysis and Critical Control Point (HACCP) system, which is designed to identify potential hazards and "prevent, reduce, or eliminate" those hazards by using processes that achieve a 5-log or 10,000-fold reduction in numbers of pathogens in the finished products.[12,14] This chapter reviews the microbiology of fruit juice and beverages in the contexts of spoilage and safety of fruit juice and beverage production.

MICROBIAL SPOILAGE OF FRUIT AND FRUIT JUICE AND BEVERAGES

BACTERIA

The most commonly encountered spoilage bacteria in fruit juices and soft drinks include species of *Acetobacter, Alicyclobacillus, Bacillus, Clostridium,*

TABLE 5.3
Bacteria Related to Spoilage in Fruit Juices and Soft Drinks

Microorganisms	Food Products	Effects
Acetobacter, Gluconobacter	Apple cider, soft drinks, fruit juice concentrate	Oxidation of ethanol, fermentation, turbidity
Lactobacillus, Leuconostoc	Orange juice concentrate, soft drinks	Sour or off-taste, buttermilk off-flavor, gummy slime or "ropiness," acetic acid, gas (CO_2), ethanol
Alicyclobacillus acidoterrestris	Apple-cranberry beverage, apple juice, orange juice concentrate, mixed fruit beverages	Phenolic or antiseptic odor or off-flavor with or without light sediment
Bacillus coagulans, B. macerans, B. polymyxa, B. licheniformis, B. subtilis	Tomato juice, soft drinks	Flat sour spoilage
Clostridium pasteurianum, C. butyricum	Tomato juice, soft drinks, fruit juice	Increased acidity, gas, strong butyric odor
Zymomonas, Saccharobacter fermentatus, Zymobacter	Apple cider, agave leaf juice	Ethanol production

Source: From Vasavada, P.C. and Heperkan, D., *Food Safety Magazine*, 8(1): 8,10,13,46–47, 2002. With permission.

Gluconobacter, Lactobacillus, Leuconostoc, Saccharobacter, Zymomonas, and *Zymobacter*.[15–23] (See Table 5.3.) Strictly aerobic, acidophilic bacteria such as *Acetobacter* and *Gluconobacter* (*Acetomonas*) have been known to cause spoilage of fruit concentrates, apple cider, and soft drinks.[15] The lactic acid bacteria, *Lactobacillus* and *Leuconostoc* spp., are also known to be associated with spoilage of fruit juice including abnormal fermentation and gas production, development of slime or ropiness, production of buttermilk-like off-flavor, and formation of cloudiness and turbidity.[22] They are among the most significant microorganisms in processing citrus juices. Spore-forming organisms (*Bacillus* and *Clostridium* spp.) are also known to cause spoilage in fruit juice and beverages.[22] Spoilage of fruit juices by *Clostridium* spp. is characterized by production of gas, a strong butyric odor, and increased acidity.[19]

Recently, *Alicyclobacillus*, an acidophilic, heat-resistant, spore-forming organism, has caused concern in the fruit juice industry.[20,24–26] The organism was first reported in 1982 as causing spoilage in apple juice in Germany.[24] The characteristic spoilage involves the formation of a phenolic or antiseptic odor with or without cloudiness and generally without gas production.[17,19]

The characteristic off-flavor associated with *Alicyclobacillus* spoilage involves guaiacol 2,6-dibromophenol and 2-methoxyphenol.[17,18,26] However, in a study of apple juice inoculated with *A. acidoterrestris*, the guaiacol content in apple juice did not always correlate with the number of cells.[27] *Alicyclobacillus* spoilage occurs seasonally, typically in the spring or summer, and occurs most commonly in apple juice and orange juice.[20,25,28] Contamination of fruit juices by *Alicyclobacillus* occurs via soil during the harvest.[28] Wisse and Parrish[27] found acidophilic, heat-resistant bacilli in the environment at one citrus processing plant. Strains of acidophilic, heat-resistant bacilli were detected in seven of 18 soil samples from orange groves, on surfaces of unwashed oranges at eight of 10 processing plants, on surfaces of washed oranges at six of nine processing plants, and in condensate water used to wash fruit at six of seven test facilities.[28] Two pear juice concentrates from 210 l drums, as well as retail packages of pear juice and orange juice nectar, also contained acidophilic, heat-resistant bacilli.[28] The researchers suggested that because fruit surfaces may be continuously contaminated with spores from the condensate wash water, the extracted juice could very well contain spores, and theoretically, contaminate the evaporator. Another study in 1999 reported finding *Alicyclobacillus* in 11/75 (14.7%) samples of concentrated orange juice.[21] In 1998, Splittstoesser et al.[25] reported that white grape and tomato juices also are susceptible to spoilage by this bacterium. In a survey of the food industry, 35% of respondents had experienced spoilage of their products due to acidophilic spore-forming bacteria. In addition to apple and orange juices, this organism also has been found in apple-grape-raspberry and apple-pear juice blend beverages.[20,28]

The genus *Alicyclobacillus* is comprised of three species: *A. acidocaldarius, A. acidoterrestris,* and *A. cycloheptanicus*. *Alicyclobacillus* spores are very heat resistant, with reported D-values ranging from 14 to 54 minutes at 90–91°C and z-values between 6 and 10°C. These bacteria can easily survive the typical heat treatment normally applied to pasteurize fruit juices.[21,29] The elevated heat resistance shown by *Alicyclobacillus* spores represents a potential risk for the deterioration of pasteurized, ultra-high temperature, or hot-fill orange juices when stored without refrigeration because the spores of this organism are able to germinate and grow at temperatures below 35°C. Growth of *Alicyclobacillus* was obtained over a pH range of 3.0 to 6.0 in an agar medium.[17] However, growth was inhibited when the ethanol concentration exceeded 6% and the sugar content exceeded 18 Brix.[16] Raising the sugar content of juices appears to increase the heat resistance of the bacteria. These results indicate that it would be more difficult to destroy the spores in a juice concentrate, as compared with a single-strength juice. Additional research has indicated that the complete elimination of these heat-resistant acidophilic bacteria from fruit juices would be difficult, but that improvement of fruit cleaning operations and condensate water systems may reduce the incidence of thermoacidophilic bacilli in fruit juices.[28] Rinsing the sanitary surfaces of equipment and evaporators with condensate water containing spores of heat-resistant bacilli may contaminate the juice entering the evaporator or the final product. The study also suggested that heat treatment in the evaporator was not sufficient to kill the spores of these bacteria.

YEASTS AND MOLDS

Yeasts and molds are major causes of spoilage of fruit juices and beverages. Yeasts predominate in the spoilage flora of fruit products because of their high acid tolerance and the ability of many of them to grow anaerobically. Reportedly, 40% of commercial fruit juices are contaminated with yeasts.[22,31–36] A high level of yeast contamination in fruit juices and soft drinks may be indicative of poor plant hygiene. Most spoilage yeasts are highly fermentative, forming ethanol and CO_2 from sugar, causing split cans and cartons, and explosions in glass or plastic bottles.[22] Pitt and Hocking[31] have listed yeasts predominantly responsible for spoilage of fruit juice, concentrates, and soft drinks including *Brettanomyces intermedius, Saccharomyces bailii, S. bisphorus, S. cerevisiae, S. rouxii, Schizosaccharomyces pombe,* and *Torulopsis holmii.* Parish and Higgins[35] isolated several species of yeast, including *Candida maltosa, Candida sake, Hanseniaspora guilliermondii, Hanseniaspora sp., Pichia membranaefaciens, Saccharomyces cerevisiae,* and *Schwanniomyces occidentalis,* from commercially produced unpasteurized orange juice. Other common yeasts include *Dekkera bruxellensis, Saccharomyces bayanus, Torulaspora delbruckii, Zygosaccharomyces microellipsodes,* and *Dekkera naardenensis (Brettanomyces naardenensis).*[22,31,32]

Mold contamination is generally not a problem in freshly squeezed orange juice unless moldy or decomposed fruit is used.[36,38] However, aerobic molds can contaminate the product, grow near the surface, and cause spoilage of fruits and soft drinks. Mold growth can result in an off-flavor or odor that may be described as "stale" or "old,"[36] development of a mycelial mat,[31] reduction in sugar content,[36] and mycotoxin production in fruit juices and soft drinks.[39,40] Heat-resistant genera of molds causing spoilage of soft drinks and fruit juices include *Byssochlamys, Paecilomyces, Neosartorya, Talaromyces,* and some species of *Eupenicillium.*[23,33,37] (See Table 5.4.) Up to 27% of samples of mango and tomato juice were reported to contain heat-resistant molds.[37] Parish and Higgins[35] isolated genera of *Aureobasidium, Cladosporium,* and *Penicillium* from pasteurized orange juice.

TABLE 5.4
Heat-Resistant Molds Isolated from Fruit Juices, Concentrates, and Soft Drinks

Product	Heat-Resistant Mold
Apple juice	*Byssochlamys fulva*
	Paecilomyces fulvus
	Talaromyces macrosporus
	Byssochlamys nivea
	Neosartorya fischeri
	Eupenicillium brefaldianum
	Talaromyces macrosporus
	Phialophora sp.
Apple concentrate	*Paecilomyces fulvus*
Apricot juice	*Byssochlamys nivea*
Berry juice	*Byssochlamys fulva*
	Eupenicillium lapidosum
Grape juice	*Byssochlamys fulva*
	Paecilomyces fulvus
	Talaromyces macrosporus
	Byssochlamys nivea
	Monascus purpureus
	Neosartorya fischeri
	Thermoascus aurannthiacum
Grape concentrate	*Byssochlamys fulva*
	Byssochlamys nivea
Pineapple juice	*Talaromyces macrosporus*
Pineapple concentrate	*Byssochlamys fulva*
	Neosartorya fischeri
	Talaromyces macrosporus
Fruit punch	*Byssochlamys fulva*
Fruit punch concentrates	*Byssochlamys nivea*
	Byssochlamys fulva
Mango concentrate	*Neosartorya fischeri*

Source: From Vasavada, P.C. and Heperkan, D., *Food Safety Magazine*, 8(1): 8,10,13,46–47, 2002. With permission.

INDICATOR BACTERIA AND PATHOGENIC ORGANISMS

Coliforms, *E. coli,* and enterococci have been isolated from citrus and other fruit products, including "fresh" (unpasteurized) juice. The coliforms may be part of the normal flora of processing plants and are not necessarily indicative of unhygienic production and processing practices. However, the presence of *E. coli* may indicate fecal contamination of the fruit surface or unsanitary handling, storage, and processing of fruit. Many pathogens readily adapt to the high-acid, low-pH juice environment and pose a public health threat.[41,42] They do not grow under acidic conditions but may survive for extended periods of time at refrigeration temperatures. Several opportunistic bacteria and yeasts such as *Klebsiella, Enterobacter, Candida,* and *Torulopsis* are often found in fruit juices. While they are very unlikely to affect healthy individuals, they are of concern to the at-risk population of immunocompromised patients, including those undergoing chemotherapy and radiation treatments.[42]

Protozoa

Cryptosporidium parvum is a significant cause of severe gastrointestinal disease in both immunocompetent and immunodeficient individuals. In 1993 and 1996, apple cider was associated with cryptosporidiosis outbreaks in which 191 people were affected (Table 5.5). It was believed that apples used for cider were contaminated when they fell on ground grazed by cattle shedding *C. parvum* oocysts or when they were washed with contaminated well water.[43,44] Deng and Cliver[45] suggested that heating for 10 to 20 sec at 70 and 71.7°C caused oocyst killing of at least 4.1 log, whereas oocyst inactivation after pasteurization for 5 sec at either temperature was 3.0 and 4.8 log, respectively. They concluded that current practices of flash pasteurization in the juice industry are sufficient to inactivate contaminant oocysts. Deng and Cliver[46] compared various methods for the detection of *C. parvum* oocysts from apple juice and found that the highest sensitivity, 10 to 30 oocysts per 100 ml of apple juice, was achieved by direct immunofluorescence assay (DIFA) followed by immunomagnetic capture (IC) of oocysts from samples concentrated by the flotation method and acid fast staining (AFS), and the polymerase chain reaction (PCR).

Pathogenic Yeasts

In addition to pathogenic bacteria, several new pathogenic yeasts, including *Candida famata (Debaryomyces hansenii), Candida guillermondii (Pichia guillermondii), Candida krusei (Issatchenkia orientalis), Candida parapsilosis,* and *Saccharomyces cerevisiae* can cause spoilage of fruit juices and beverages. These new pathogens are very unlikely to affect healthy individuals but are of concern in immunocompromised patients.[23]

Viruses

Viruses are not very common in fruit juices and products. However, contamination by hepatitis A and Norwalk-like virus (small round structured viruses,

TABLE 5.5
Microorganisms Related to Foodborne Illness in Fruit Juices

Microorganisms	Food Product	No. of Cases	Year/Country	Ref.
Bacillus cereus	Orange juice	85	1994/U.S.	44
Cryptosporidium parvum	Apple cider (unpasteurized)	31	1996/U.S.	44
Cryptosporidium parvum	Apple cider (unpasteurized)	160	1993/U.S.	43, 44
Cryptosporidium parvum	Apple juice	NS	?	66
E. coli O157:H7	Apple cider	9	1999/U.S.	44
E. coli O157:H7	Apple cider	13	1991/U.S.	63
E. coli O157:H7	Apple cider	14	1980/Canada	44
E. coli O157:H7	Apple juice (unpasteurized)	70	1996/U.S. and Canada	44, 63
E. coli O157:H7	Apple juice (unpasteurized)	6	1996/U.S.	13, 44
E. coli O157:H7	Apple juice	10	1996/U.S.	44
E. coli O134	Orange juice	NS	?	65
Salmonella anatum	Orange juice	4	1999/U.S.	44
S. enteritidis	Orange juice	74	2000/U.S.	67
S. hartford *S. gaminara* *S. rubislaw*	Orange juice	62	1995/U.S.	44
S. muenchen	Orange juice	220	1999/U.S. and Canada	44
S. typhi	Frozen mamey (often used to make juice)	16	1999/U.S.	44
S. typhi	Orange juice	44	1989/U.S.	44
S. typhimurium	Apple cider	~300	1974/U.S.	44
	Orange juice	427	1999/Australia	64
Norwalk-like virus	Fruit smoothies	24	2000/U.S.	44
Small round structured viruses (SRSVs)	Orange juice	3000	?	23

NS = not specified

SRSVs) has been reported.[23,44] In April 2000, 24 people attending a conference in Atlanta suffered from viral gastroenteritis associated with fresh squeezed unpasteurized fruit smoothies. Norwalk-like virus was detected in three stool samples from patients suffering from the illness.[47]

Mycotoxins

Several species of molds are capable of producing different mycotoxins in fruit juices. Mycotoxins, particularly patulin, represent a potent food safety hazard in fruit juice and beverages. Some molds, e.g., *Penicillium expansum, P. griseofulvum, P. roqueforti* var. *carneum, P. funiculosum, P. claviforme, P. granulatum*,[39,40,48,49] and *Byssochlamys* spp.,[39,40] produce patulin in apple juice, while others such as *Neosartorya* produce fumitremorgins, terrein, verruculogen, and fischerin. *Byssochlamys* species also produce byssotoxin A and byssochlamic acid.[40]

Mycotoxin production in fruit juice is a global problem. Patulin production in fruit juice has been reported in several countries: 65% of 113 samples of apple juice in Australia,[50] 44% of 215 samples of apple juice concentrates in Turkey,[51,52] 3% of 111 samples of processed apple and grape juice in Brazil, and 23% of 40 apple juice samples in the U.S. tested positive for patulin.[53] Patulin in apple juice may be eliminated by fermentation of the apple juice to cider or addition of ascorbic acid.[53] Other mycotoxins produced in fruit juice by molds include ochratoxin A, citrinin, and penicillic acid. Caffeine inhibits aflatoxin production[54] but does not inhibit ochratoxin A found after mold growth on coffee beans.[55,56]

EMERGING PATHOGENS AND OUTBREAKS OF ILLNESS

Although fruit juices have been recognized as vehicles of foodborne illness since 1922, pathogenic organisms were not considered a major cause for concern in fruit juices and fruit beverages until recently.[57] Despite the occasional reports of foodborne illness outbreaks from consumption of apple and orange juices and despite documented evidence of the ability of some pathogens to survive in fruit juices, most low-pH, high-acid foods were not considered potentially hazardous foods. However, an unprecedented rise in the number of foodborne illness outbreaks, consumer illness associated with juice products, and recalls of fruit juice and juice products during the past decade have led to a recognition of emerging pathogens as a major threat to the safety of fruit juice and beverages. The following is a brief review of some of the well-known outbreaks and major emerging pathogens.

Early Outbreaks

There have been at least eight outbreaks of illness from consumption of commercial "fresh" or "unpasteurized" fruit juices since 1922, when apple cider was implicated in an outbreak of typhoid fever.[57] Since that time, outbreaks from fruit juice consumption have occasionally been reported

(Table 5.5). In 1944, juice contaminated by an asymptomatic food handler resulted in 18 cases of typhoid fever and one death in Cleveland.[57] In 1966, an outbreak of gastroenteritis was reported at a university in which a causative agent was not found but a frozen orange juice product was implicated as the source of the illness.[57] A 1967 outbreak from contaminated water added to orange juice concentrate made 5200 people ill and was caused by a virus.[57,58] In a 1974 outbreak in New Jersey, about 300 people reportedly became ill from *Salmonella typhimurium* in apple cider. Some of the apples used in the cider manufacturing had been picked up from the ground in an orchard fertilized with manure.[8] Manure is suspected to be the cause of outbreaks involving several pathogens, including *E. coli* O157:H7. In 1980, 13 or 14 children in Canada were reported suffering from a serious illness associated with consumption of fresh apple juice.[12] The stool samples from children affected by the illness tested negative for major known human pathogens, including enteropathogenic *E. coli*, *Salmonella*, *Shigella*, *Campylobacter*, and *Yersinia*.[12] This was probably the first reported incidence of bloody diarrhea and the hemolytic uremic syndrome (HUS) associated with infection with *E. coli* O157:H7 in Canada but was not reported as such because the organism was not recognized as a human pathogen until 1982.[60]

A 1989 outbreak in a New York hotel was caused by orange juice contaminated with *Salmonella typhi*. In this outbreak, 45 confirmed cases and 24 probable cases of typhoid fever with 21 hospitalizations were reported.[57] The outbreak was attributed to an asymptomatic food worker who contaminated the product during the reconstitution of concentrated orange juice.[57] Orange juice products have also been involved in hepatitis A and gastroenteritis outbreaks.[58,59,74]

OUTBREAKS IN THE 1990S

Recently, an increase has been observed in foodborne illness linked to fresh fruits and vegetables, and juice and cider products contaminated with so-called emerging pathogens such as *Escherichia coli* O157:H7, *Salmonella* spp., and *Cryptosporidium* spp. (Table 5.6).[44] Enterohemorrhagic *E. coli* (EHEC) has been the emerging pathogen most frequently isolated in outbreaks associated with unpasteurized juice and cider since 1990 (Table 5.6).

Two multistate outbreaks of salmonellosis associated with fruits and vegetables occurred in 1990.[67–69] According to the U.S. government's Centers for Disease Control and Prevention (CDC), *Salmonella chester* associated with cantaloupes affected 245 persons in 30 states, and *Salmonella javiana* associated with tomatoes affected 174 persons in four states.[67,68] The illness was associated with consumption of contaminated cantaloupes in fruit salad and from salad bars.

TABLE 5.6
Reported Foodborne Outbreaks Linked to Unpasteurized Juice/Cider Since 1990

Year	Product	Pathogen	Location	No. of Cases
1991	Apple cider	E. coli O157:H7	Massachusetts	23
1993	Apple cider	Cryptosporidium parvum	Maine	160
1995	Orange juice	Salmonella spp.	Florida	63
1996	Apple cider	E. coli O157:H7	Connecticut	10
1996[a]	Apple cider	E. coli O157:H7	Western U.S. and Canada	66
1996	Apple cider	E. coli O157:H7	Washington	2
1996	Apple cider	Cryptosporidium parvum	New York	31
1998[b]	Apple cider	E. coli O157:H7	Ontario	14
1999	Orange juice	Salmonella typhimurium	Australia	400
1999	Orange juice	Salmonella muenchen	Arizona, Western U.S., and Canada	423
1999	Apple cider	E. coli O157:H7	Oklahoma	9
2000	Orange juice	Salmonella enteritidis	Arizona, California, Colorado, Minnesota, Nevada, Wyoming, Washington	143
2000	Unpasteurized fruit smoothies	Viral gastroenteritis Norwalk-like virus	Georgia	24

[a] Unpasteurized juice from California was involved. One child died in the U.S.
[b] Local health officials identified one batch of noncommercial, custom-pressed apple cider as the most likely source.

A multistate outbreak of *Salmonella poona* infections affecting more than 400 persons in 23 states in the U.S. and in Canada was reported in 1991.[70] Also in 1991, an outbreak of *E. coli* O157:H7 occurred in Massachusetts in which the implicated food vehicle was fresh-pressed unpasteurized apple juice.[71] In this outbreak, 23 individuals had diarrhea, 16 had bloody diarrhea, and four developed HUS.

In 1995, a first documented outbreak of salmonellosis occurred in Florida in which a citrus processing facility was implicated.[72] Sixty-two confirmed and probable cases of salmonellosis were reported, although CDC estimated the total number of cases to be between 630 and 6300. *Salmonella* serovars *hartford, rubislaw, saintpaul, newport* and *ganminara* were isolated from clinical samples, the plant environment, and amphibians (toads) collected from near the processing plant.

In October 1996, unpasteurized apple cider or juice was associated with three outbreaks of illness. An outbreak of *E. coli* O157:H7 infections associated with an unpasteurized commercial apple juice caused 66 illnesses and one death.[9] This outbreak resulted in a nationwide recall of all products from the company and an eventual multimillion dollar settlement. A small outbreak of *E. coli* O157:H7 occurred in October 1996 in Connecticut in which 14 persons were affected after drinking apple cider.[10] Seven were hospitalized, three with HUS and one with thrombotic thrombocytopenic purpura (TTP). The illness was associated with a particular brand of cider pressed at a mill where some of the apples used were "drop" apples. The cider was not pasteurized. Also in 1996, another small outbreak of *E. coli* O157:H7 illness affecting six people occurred in Washington state.[73] The apple cider was made at a church event. The apples were washed in chlorine, but the concentration of chlorine was not known.[73]

Cryptosporidium is another emerging pathogen associated with at least three outbreaks related to drinking apple cider. In 1993, two outbreaks of cryptosporidiosis occurred, one in Maine and the other in New York state.[10] In the first outbreak, the apples used for cider came from trees near a cow pasture. In the second case, the rinse water used came from a well contaminated with coliforms. In another cryptosporidiosis outbreak in New York state, 21 confirmed and 11 suspect cases were reported after drinking apple cider produced at a local cider mill located across the road from a dairy farm.[43] While testing of cider samples, equipment swabs, and well water did not yield *Cryptosporidium*, coliforms were detected in well water samples, and at least one well water sample tested positive for *E. coli*.

During 1999, more than 300 cases of diarrheal illness due to *Salmonella muenchen* were reported in Washington, Oregon, and Canada.[12] These cases were attributed to commercially distributed unpasteurized apple juice produced by a single processor in Tempe, Arizona. In Washington state, the outbreak was linked to restaurant patrons drinking a fruit smoothie containing unpasteurized juice or eating in an establishment where the juice was served. In Oregon, the ill persons were among those who had eaten a buffet brunch and drunk the unpasteurized juice produced by Sun Orchard.[12] In addition to Washington and Oregon, cases of salmonellosis were reported in 13 other states including Arizona, California, Connecticut, Florida, Illinois, Iowa, Massachusetts, Michigan, Minnesota, New Mexico, Texas, Utah, and Wisconsin, as well as in two Canadian provinces, Alberta and British Columbia. This was the second largest *Salmonella* outbreak associated with unpasteurized orange juice.[22] As of April 2000, a total of 423 cases, including one fatality, from *S. muenchen* had been reported.[12]

Also in 1999, an outbreak of illness linked to *E. coli* O157:H7 in commercially processed unpasteurized apple juice was reported in Oklahoma. This outbreak involved nine cases, including seven children, six hospitalizations and four HUS cases.[15]

Unpasteurized orange juice was associated with an outbreak of illness due to *Salmonella enteritidis* in April 2000.[14] By May 2000, 143 cases attributed to this contaminated juice were reported in Arizona, California, Colorado, Minnesota, Nevada, Washington, and Wyoming.[15]

A multistate outbreak of infection associated with the emerging pathogen *Cyclospora cayetanensis* occurred in 1996.[62,69] Approximately 850 cases of laboratory-confirmed *Cyclospora* infections were reported in ten states in the U.S. and in Ontario and were traced to consumption of raspberries imported from Guatemala. Unlike *Cryptosporidium* and other parasitic diseases, most reported cases of *Cyclospora* infection occurred in immunocompetent patients. No deaths were reported.

ILLNESS FROM OTHER POTENTIAL FOOD SAFETY HAZARDS

In addition to emerging pathogenic bacteria, fruit juice and beverages may be contaminated by pathogenic yeasts, molds, mycotoxins, and metal ions—all of which are potential food hazards. Outbreaks of illness and recalls have been prompted by contamination of juice and products by tin, lead, residues of cleaning chemicals, pieces of glass and plastic, etc.[13,15,44] (Table 5.7 and Table 5.8). These hazards are not controlled by heat or pasteurization and must be addressed through implementation of a well-designed HACCP plan.

TABLE 5.7
Injury or Illness Associated with Chemical and Physical Hazards in Fruit Juices

Year	Food Product	Hazard	No. of Cases	Location
1997	Pineapple juice	Tin	19	Texas, Florida
1992	Canned fruit nectar	Lead	1	California
1992	Fruit drink	Lead cleaning solution residue	3	New York, New Jersey, Vermont
1990	Guanabana juice	Toxic seed material	9	Texas
1983	Elderberry juice	Poisonous plant parts	11	California
1969	Tomato juice	Tin	113	Washington, Oregon

TABLE 5.8
Recalls Due to Contaminated Fruit Juices

Year	Food Product	Hazard	Location
2001	Orange juice	Mold contamination	Nationwide
2000	Apple juice	Fermentation and off-taste due to *Lactobacilli* contamination	New York, New Jersey, Pennsylvania
2000	Apple juice	Mold contamination	Nationwide
2000	Apple juice	Yeast contamination	Eight states
2000	Citrus juice	Possible *Salmonella* contamination	Six states
2000	Orange fruit drink	Sour off-taste	Florida, Georgia, South Carolina
1999	Apple juice	Small pieces of glass	N/A
1999	Orange juice	Fermentation	Nationwide, Canada
1999	Cranberry–raspberry drink	Mold contamination	Southeastern U.S.
1998	Raspberry drink	Mold contamination	Ten states
1998	Orange juice	*Salmonella* contamination	Eight states and one Canadian province
1997	Orange juice	Glass	Multistate
1997	Citrus beverages	Plastic	Multistate
1996	Infant apple–prune and prune juice	Lead	Multistate
1994	Orange juice	Fermentation, *Bacillus cereus*, and yeast	Mobile County, Alabama
1991	Fruit punch	Glass	New York
1991	Citrus punch drink	Cleaner residue	New Jersey, Delaware, Pennsylvania, New Hampshire
1988	Fruit punch	Tin	California

ENSURING SAFETY OF JUICE: STRATEGY AND CONTROL

In the aftermath of the 1996 outbreaks of *E. coli* O157:H7, FDA held a public meeting to provide for information exchange on current industry practices for the production of juice products, review the risk associated with emerging pathogens and other hazards in fresh juice and products, discuss possible scientific and technological solutions, identify the areas for research, and consider measures necessary to improve the safety of fruit juice and products.[13] This meeting and subsequent deliberations led

Microbiology of Fruit Juice and Beverages

to current strategy and approaches for ensuring safety of fruit juice. The main strategy of controlling microbial pathogens and other hazards includes pasteurization of juice and products, design and implementation of HACCP, and establishing current Good Manufacturing Practices (cGMPs) in juice manufacturing. The Fresh Produce Subcommittee (FPS) of the National Advisory Committee on Microbiological Criteria for Foods (NACMCF) attended the open public meeting, deliberated issues raised at the meeting, and made specific recommendations to the NACMCF.[12] The NACMCF stated that many aspects affect pathogen control such as agricultural practices; product handling; equipment used; growing location, including produce obtained from below ground (carrots), on ground (e.g., drops), or from trees; pH; acidulants; method of processing; degree of animal contact; refrigeration; packaging; and the distribution system. The NACMCF reached the following conclusions:

- While the risks associated with specific juices vary, safety concerns are associated with juices, especially unpasteurized juices.
- The history of public health problems associated with fresh juices indicates a need for active safety interventions.
- For some fruit, intervention may be limited to surface treatment, but for others, additional interventions may be required (e.g. pasteurization).

The NACMCF recommended[75,80] to FDA the use of safety performance criteria instead of mandating the use of a specific intervention technology. In the absence of specific pathogen–product associations, the committee recommends the use of *Escherichia coli* O157:H7 or *Listeria monocytogenes* as the target organism, as appropriate. The NACMCF believed that a tolerable level of risk may be achieved by requiring an intervention(s) that has been validated to achieve a cumulative 5-log reduction in the target pathogen(s) or a reduction in yearly risk of illness to less than 10^{-5}, assuming consumption of 100 ml of juice daily.

The NACMCF stated that HACCP and safety performance criteria should form the general conceptual framework needed to ensure the safety of juices, that control measures should be based on a thorough hazard analysis, and that validation of the process must be an integral part of this framework. The NACMCF recommended mandatory HACCP for all juice products and that processing plants should implement and strictly adhere to industry cGMPs. In addition, the NACMCF recommended industry education programs addressing basic food microbiology, the principles of cleaning and sanitizing equipment, GMPs, and HACCP.[75]

The NACMCF recommended further study in the following areas:

- Research on the efficacy of new technologies and intervention strategies for safety
- Research on the contamination, survival, and growth of pathogens on produce with or without breaks in the skin or areas of rot, and within the core
- Research on how produce becomes contaminated with human pathogens including the relevant microbial ecology during production and processing of juice (In particular, there is an urgent need for these types of studies on *E. coli* O157:H7 in apple juice.)
- Baseline studies on the incidence of human pathogens on fruits and vegetables, particularly those used in juice processing

Research on labeling information needed for consumer understanding and choice of safer juices and juice products

The NACMCF agreed that there is a need to understand the differences among various juices and juice products (e.g., citrus vs. other) and noted that consumers presently do not have a means to clearly differentiate between unpasteurized and pasteurized products. Terms used to refer to juice products do not always have universal meanings, e.g., "cider" is perceived to be an unpasteurized product whereas "juice" is often perceived to be pasteurized. The NACMCF stated that traditional heat treatments given to juices and juice products have been designed to achieve shelf stability, to remove water (i.e., concentration), or to affect other quality-related factors. These treatments, commonly referred to as "pasteurization," are greatly in excess of a process needed to inactivate foodborne pathogens.

The NACMCF stated that it could not strongly endorse labeling as an interim safety measure because of the lack of sufficient data to evaluate the effectiveness of labeling statements for safety interventions or to inform consumer choice. However, as an interim safety measure, the FDA decided to require a labeling statement (Figure 5.2) for packaged juice products not specifically processed to eliminate harmful bacteria.[14]

In response to the 1997 presidential directive designed to "ensure the safety of imported and domestic fruits and vegetables" and "to provide further assurance that fruits and vegetables consumed by Americans meet the highest health and safety standards," the FDA and the U.S. Department of Agriculture (USDA) issued a document titled *Guidance for Industry: Guide to Minimize Microbial Food Safety Hazards For Fresh Fruits and Vegetables*[76] to addresses microbiological food safety and good agricultural and management practices and to help fruit and vegetable producers ensure the safety of their produce.

Microbiology of Fruit Juice and Beverages

> **WARNING:**
> This product has not been pasteurized and therefore may contain harmful bacteria that can cause serious illness in children, the elderly, and persons with weakened immune systems.

FIGURE 5.2 Labeling statement for fresh (unpasteurized) packaged juice products.

GMP AND BEST PRACTICES FOR JUICE PROCESSORS

In order to understand the microbiological hazards and best practices for controlling these hazards, the FDA conducted a field assignment to inspect fresh, unpasteurized apple cider operations in 1997.[77] Results of 237 inspections conducted in 32 states indicated that 52% of the firms had no objectionable or minor insanitary conditions and were classified as "No Action Indicated" (NAI). Thirty-six percent of the firms were assigned "Voluntary Action Indicated" (VAI), meaning that objectionable conditions of minor significance were observed, but no administrative or regulatory follow up was required.[78] No firm was assigned "Official Action Indicated" (OAI). Based on the inspection findings, 67% of the firms were characterized as having good sanitation, 27% were marginal, and 4% had poor sanitary conditions. The FDA inspections also identified common conditions found in plants operating under good and poor sanitary conditions.[79] These are summarized in Table 5.9.

The majority of fresh apple cider operations are local or intrastate operations and hence are regulated by state agencies. Recognizing the potential food safety hazards associated with fresh apple cider and unpasteurized juice, many states have established cGMPs for the cider and juice industry.[80] These include:

Harvesting
- Avoiding use of drops, rotten fruits, or fruits soiled by birds or manure for unpasteurized apple juice
- Using only sound apples meeting standards for U.S. cider and rejecting fruits dried before dumping
- Using clean containers for harvesting and storing apples
- Applying Good Hygienic Practices, including providing readily accessible toilets and handwashing facilities for the workers

TABLE 5.9
Conditions Observed in the FDA Inspections of Fresh Unpasteurized Apple Cider Manufacturers

Typical Good Operations	Typical Poor Operations
Apples are culled at harvesting so only wholesome apples are collected.	No sanitation facilities are provided in the orchard during harvesting.
If drops are used, they are carefully managed and washed.	Domestic animals are grazing adjacent to orchard, or stabled or penned adjacent to processing facility.
All wash water comes from a protected source that is tested for microbiological indicators at least yearly just before the season starts.	Drops are used and mixed with tree-picked or otherwise poorly managed.
	Apples are inadequately culled, and badly bruised apples are used.
The apples are spray-washed and wet-brushed prior to pressing with water treated with an antimicrobial agent.	The wash water is from a nonpublic water supply source that is not sampled and tested annually and as required by state water quality regulations.
The conveyor system, chopper, and pressing equipment are of sanitary design and condition. The equipment is free of organic residues, and any hardwoods have a smooth, easily cleanable surface.	Apples are not washed or are inadequately washed prior to pressing, or flume water is not flowing or changed frequently.
Once apples are pressed, the cider is kept in closed piping and vats, is promptly cooled, and is not exposed to cross-contamination.	Equipment is in disrepair, especially wood surfaces, and is not easily cleanable.
	Food contact surfaces are not clean or are not properly cleaned and sanitized between uses.
The processing area is enclosed, clean and uncluttered, and free from flying insects.	Personal hygiene is poor and there are no handwashing facilities.
The equipment is cleaned and sanitized, and the processing area is cleaned, after processing and/or before processing is resumed.	There are open entryways from the outside. There are flying insects in the processing area, on equipment, and in vats holding cider.
Toilet facilities and handwashing facilities are available, and employees practice good hygiene with respect to handwashing, clothing and hair restraints, and eating and smoking behavior.	
The plastic, finished product containers are stored in a clean and protected area, are not reused, and are clean when filled.	

Source: From U.S. Food and Drug Administration Center for Food Safety and Applied Nutrition, Report of 1997 Inspections of Fresh Unpasteurized Apple Cider Manufacturers: Summary of Results: Analysis of Inspectional Findings, January 1999, available at http://vm.cfsan.fda.gov/~dms/ciderrpt3.html.

Receiving
- Adequate record keeping and proper storage of apples in clean, sanitary containers
- Proper inspection of apples upon receipt
- Identifying date of purchase, source of product, and type of product
- Thorough cleaning and washing of apples before crushing
- Proper storage

Processing
- Inspection of apples before washing/processing
- Discarding wormy, decayed, or rotten apples and using only intact, wholesome apples
- Washing and cleaning of apples before crushing
- Using food grade detergents and sanitizers and controlling temperature and sanitizer concentration in flume water
- Proper cleaning and sanitation of crushing and pressing equipment, tubing, press racks, press cloth, etc.
- Sanitary handling of processing and proper disposal of waste/pomace
- Proper use of additives, such as sodium benzoate and potassium sorbate
- Proper processing and sanitary bottling of the apple juice/cider
- Microbiological testing for total coliforms, fecal coliforms, and *E. coli*
- Proper labeling of retail containers, and handling, storage, and transportation of cider at <45°F

The attributes of plants typically operating under good and poor conditions and the GMPs provide a basis for so-called "Best Practices" for cider/ juice production,[78] which are listed below:

- *Culling* — Prompt and effective culling of apples after harvesting to remove cut, badly bruised, rotten, and insect- or bird-damaged apples that might have been contaminated with pathogens
- *Initial washing* — Prompt washing of apples after harvesting and culling to clean the surface and reduce the transfer of organic material from apples that may be contaminated with pathogens
- *Prompt processing or refrigerated holding* — Prompt processing of apples after harvesting to reduce the potential for pathogen growth during extended holding or refrigerated storage of apples if processing is delayed more than one day

- *Final culling, washing, and brushing* — Effective culling, washing, and wet-brushing of apples with water containing an antimicrobial agent immediately prior to crushing/chopping to remove damaged apples that may have been contaminated with pathogens and to provide a clean surface that will not introduce pathogens into the cider during processing
- *Closed processing system* — Effective containment of the cider in a system of closed pipes and covered vats to the extent possible after pressing to reduce the risk of cross-contamination from environmental sources and employee handling
- *Equipment sanitation* — Prompt and effective maintenance, cleaning, and sanitizing of all food contact surfaces on equipment including sprayer-brush units, conveyors, and any hardwood surfaces to avoid the buildup of organic residues that may harbor microorganisms as well as the proper cleaning, drying, and storage of press cloths
- *Environmental sanitation* — Maintenance of a sanitary environment by providing wall, ceiling, and floor surfaces that are easily cleanable and clean; having screened or closed entryways; eliminating flying insects; utilizing a safe water supply; promptly removing and properly storing pomace; maintaining adjacent grounds free of debris, trash, and pest harborages; and keeping domestic animals and animal pens well removed from the processing facility
- *Employee hygiene* — Maintenance of clean and functional toilet facilities, handwashing facilities, and hand sanitizing stations in the production area; and enforcement of good hygienic practices involving handwashing, protective clothing, and no eating or smoking in the production area

Model HACCP

The Hazard Analysis and Critical Control Point (HACCP) program is a systematic, proactive, and preventative approach to food safety assurance that involves identifying and assessing the microbiological, chemical, and physical hazards from a particular food production process or practice (Hazard Analysis) and minimizing the risk by controlling or eliminating the hazards at the points in the production process where a failure would likely result in a food hazard being introduced or allowed to persist (Critical Control Points).[81–83] The HACCP system has been effectively used to control microbial hazards in low-acid canned foods, milk and dairy foods, and seafoods, and more recently, in meat and poultry. Recently, the FDA published the juice HACCP final rule,[15] requiring all juices and juice ingredients in beverages to be processed under HACCP. All juice processors except very small businesses in the U.S. were required to comply with the HACCP rule by January 21, 2003. A model HACCP plan for fresh-squeezed (not pasteurized) citrus juice operation was developed by the University of Florida.[81] As in other HACCP plans, the model HACCP plan is based on seven principles:

1. Conduct a hazard analysis.
2. Identify critical control points.
3. Establish critical limits.
4. Establish monitoring/inspection requirements.
5. Establish corrective actions.
6. Establish recordkeeping system.
7. Establish verification and validation procedures.

Figure 5.3 shows a flow diagram with critical control points for fresh-squeezed (not pasteurized) citrus juice operation.[81] Of course, before developing a HACCP plan, the juice processor must have and implement the GMP requirements and Sanitation Standard Operating Procedures (SSOPs)

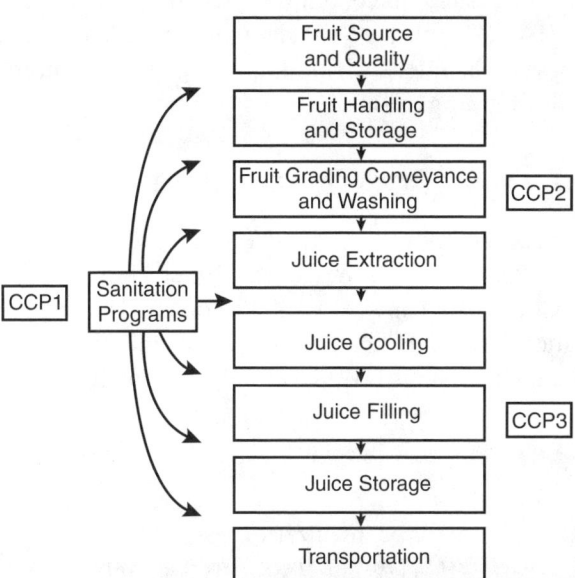

FIGURE 5.3 Flow diagram of fresh squeezed (unpasteurized) fruit juice. (From A Model HACCP Plan for Small Scale, Fresh Squeezed (Non-pasteurized) Citrus Juice Operations, CIR 1179, University of Florida Cooperative Extension, 1997.)

as prerequisite programs. The HACCP regulations also require that certain functions relating to the HACCP operation be performed by individuals trained in the application of HACCP under a standardized curriculum recognized by the FDA or by individuals having equivalent knowledge through job experience.

SUMMARY

Fruit juices and fruit-based beverages are an important segment of the domestic and international market. Microorganisms, particularly yeast and lactic acid bacteria, play a significant role in spoilage of fruit juice and beverages. Pathogenic bacteria are usually not a problem in fruit juices and beverages, but recent outbreaks of foodborne illness, at least one involving a fatality, attributed to consumption of commercial, nonpasteurized ("fresh") fruit juices contaminated with emerging pathogens such as *Escherichia coli* O157:H7, *Salmonella,* and *Cryptosporidium parvum* have caused concern among the consuming public. The FDA has issued recommendations of "best practices" for fresh juice processors, required a label for juices and juice products not treated to attain a 5-log reduction in the pertinent microorganisms, and issued a regulation for a mandatory Hazard Analysis and Critical Control Point (HACCP) system designed to ensure safety of fruit juice and juice beverages. The implementation of Best Practices and HACCP will lead to adequate control of microbial hazards and to a decrease in incidents of spoilage, recalls, and outbreaks of illness associated with microorganisms.

ACKNOWLEDGMENT

This is a contribution from the College of Agriculture, Food and Environmental Sciences, University of Wisconsin, River Falls and the Cooperative Extension Service of the University of Wisconsin. The technical assistance and collaboration of Dr. Dilek Heperkan, Istanbul Technical University, during her sabbatical visit at UW River Falls is gratefully acknowledged.

REFERENCES

1. Enright, A., Financials eclipse fizzle, *Prepared Foods*, 170(4), 41–42, 2001.
2. Prince, G.W., Great approval ratings, *Beverage World,* July 15, 1999.
3. Beverage World (www.beverageworld.com), August 2000.
4. O'Donnell, C.D., Healthy halos, *Prepared Foods,* 170(4), 50–51, 2001.
5. Roberts, W. and Dornblaser, L., Beverages: absolutely FABulous, *Prepared Foods,* 171(4), 19–29, 2002.

6. Rowles, K., Processed apple products and marketing analysis: apple juice and cider, SP 2002–01, Cornell University, Ithaca, NY, 2001.
7. Putnam, J.J. and Allshouse, J.E., Food Consumption, Prices and Expenditure: 1970–1997, Food and Rural Economics Division, USDA Economic Research Service, Statistical Bulletin No. 965, April 1999.
8. Centers for Disease Control and Prevention (CDC), *Salmonella typhimurium* outbreak traced to a commercial apple cider — New Jersey, *MMWR*, 24, 87–88, 1975.
9. Centers for Disease Control and Prevention (CDC), Outbreak of *Escherichia coli* O157:H7 infections associated with drinking unpasteurized commercial apple juice — British Columbia, California, Colorado, and Washington, *MMWR*, 45, 975, 1996.
10. Centers for Disease Control and Prevention (CDC), Outbreaks of *Escherichia coli* O157:H7 infections and cryptosporidiosis associated with drinking unpasteurized apple cider — Connecticut and New York, *MMWR*, 46, 4–8, 1997.
11. Parish, M.E., Public health and nonpasteurized fruit juices, *Crit. Rev. Microbiol.*, 23, 109–119, 1997.
12. Centers for Disease Control and Prevention (CDC), Outbreaks of *Salmonella* serotype *muenchen* infections associated with unpasteurized orange juice — United States and Canada, *MMWR*, 48, 581–585, 1999.
13. Food and Drug Administration (FDA), Hazard analysis and critical control point (HACCP); procedures for the safe and sanitary processing and importing of juice, *Federal Register*, 63, 20450–20486, 1998.
14. Food and Drug Administration (FDA), Food labeling: Warning and notice statement: Labeling of juice products; Final rule, *Federal Register*, 66, 37029–37056, 2001.
15. Food and Drug Administration (FDA), Hazard analysis and critical control point (HACCP); procedures for the safe and sanitary processing and importing of juice; final rule, *Federal Register*, 66, 6138–6202, 2001.
16. Baird-Parker, T.C. and Kooiman, W.J., Soft drinks, fruit juices, concentrates, and fruit preserves, 643–668, ICMSF, International Commission on Microbiological Specifications for Foods of the International Association of Microbiological Societies in Microbial Ecology of Foods, Academic Press, London, 1980.
17. Splittstoesser, D.F., Churey, J.J., and Lee, C.Y., Growth characteristics of aciduric sporeforming bacilli isolated from fruit juices, *J. Food Prot.*, 57, 1080–1083, 1994.
18. Pettipher, G.L., Osmundson, M.E., and Murphy, J.M., Methods for the detection and enumeration of *Alicyclobacillus acidoterrestris* and investigation of growth and production of taint in fruit juice and fruit juice–containing drinks, *Lett. Appl. Microbiol.*, 24, 185–189, 1997.
19. Morton, R.D., Spoilage of acid products by butyric acid anaerobes — a review, *Dairy Food Environ. Sanit.*, 18(9), 580–584, 1998.
20. Walls, I. and Chuyate, R., *Alicyclobacillus* historical perspective and preliminary characterization study, *Dairy Food Environ. Sanit.*, 18, 499–503, 1998.
21. Eiroa, M.N.U., Junqueira, V.C.A., and Schmidt, F.L., *Alicyclobacillus* in orange juice: occurrence and heat resistance of spores, *J. Food Prot.*, 62, 883–886, 1999.

22. Jay, M.J., Fruit and vegetable products: whole, fresh cut, and fermented, in *Modern Food Microbiology,* Aspen Publishing, Gaithersburg, MD, 2000, pp. 151–152.
23. Stratford, M., Hofman, P.D., and Cole, M.B., Fruit juices, fruit drinks, and soft drinks, in *The Microbiological Safety and Quality of Food,* Lund, B.M., Baird-Parker, T.C., and Gould, G.M., Eds. Aspen Publishing, Gaithersburg, MD, 2000, pp. 836–869.
24. Cerny, G., Hennlich, W., and Poralla, K., Fruchtsaftverderb durch Bacillen: Isolierung und Charcterisierung des Verderbserregers, *Z. Lebensm. Unters. Forsch.,* 179, 224–227, 1984.
25. Splittstoesser, D.F, Lee, C.Y., and Churey, J.J., Control of *Alicyclobacillus* in the juice industry, *Dairy Food Environ. Sanit.,* 18, 585–587, 1998.
26. Baumgart, J., Huessmann, M., and Schmidt, C., *Alicyclobacillus acidoterrestris:* occurrence, significance and detection in beverages and beverage base, *Flussiges Obst.,* 64, 178–180, 1997.
27. Orr, R.V., Shewfelt, R.L., Huang, C.J., Tefera, S., and Beuchat, L.R., Detection of guaiacol produced by *Alicyclobacillus acidoterrestris* in apple juice by sensory and chromatographic analyses, and comparison with spore and vegetative cell populations, *J. Food Prot.,* 63, 1517–1522, 2000.
28. Wisse, C.A. and Parrish, M.E., Isolation and enumeration of sporeforming thermoacidophilic, rod-shaped bacteria from citrus processing environments, *Dairy Food Environ. Sanit.,* 18, 504–509, 1988.
29. Palop, A., Alvarez, I., Raso, J., and Condon, S., Heat resistance of *Alicyclobacillus acidocaldarius* in water, various buffers, and orange juice, *J. Food Prot.,* 63, 1377–1380, 2000.
30. Vasavada, P.C. and Heperkan, D., Non-thermal alternative processing technologies for the control of spoilage bacteria in fruit juices and fruit-based drinks, *Food Safety Magazine,* (8)1: 8,10,13,46–47, 2002.
31. Pitt, J.I. and Hocking, A.D., *Fungi and Food Spoilage,* Academic Press, Sydney, 1985.
32. Walker, H.V. and Ayres, J.C., Yeasts as spoilage organisms, in *The Yeasts, Vol. 3, Yeast Technology,* Rose, A.H. and Harrison, J.S., Eds., Academic Press, London, 1970, pp. 464–527.
33. Beuchat, L.R., Spoilage of acid products by heat resistant molds. *Dairy Food Environ. Sanit.,* 18, 588–593, 1998.
34. Splittstoesser, D.F. and King, D., in *Compendium of Methods for the Microbiological Examination of Foods,* 2nd ed., Speck, M., Ed., American Public Health Association, Washington, DC, 1984.
35. Parish, M. and Higgins, D.P., Yeast and molds isolated from spoiling citrus products and by-products, *J. Food Prot.,* 52, 261–263, 1988.
36. Murdock, D.I., Microbiology of citrus products, in *Citrus Science and Technology,* Vol. 2., AVI Publishing, Westport, CT, 1977.
37. Obeta, J.A.N. and Ugquanyi, J.O., Heat-resistant fungi in Nigerian heat-processed fruit juices, *Int. Food Sci. Technol.,* 30, 587–590, 1995.
38. Eleftheriadou, M., et al., Factors affecting quality and safety of freshly squeezed orange juice, *Dairy Food Environ. Sanit.,* 18, 14–23, 1998.

39. Scott, P.M., et al., Occurrence of patulin in apple juice, *J. Agric. Food Chem.,* 20, 450–451, 1972.
40. Rice, S.L., Beuchat, L.R., and Worthington, R.E., Patulin production by *Byssochlamys* spp. in fruit juices, *Appl. Environ. Microbiol.,* 34, 791–796, 1977.
41. Narciso, J.A. and Parish, M.E., Relationship of molds in paperboard packaging to food spoilage, *Dairy Food Environ. Sanit.,* 20, 944–951, 2000.
42. Lund, B.M. and Snowdon, A.L., Fresh and processed fruits, in *The Microbiological Safety and Quality of Food,* Lund, B.M., Baird-Parker, T.C., and Gould, G.M., Eds., Aspen Publishing, Gaithersburg, MD, 2000, pp. 738–758.
43. Millard, P.S., Gensheimer, K.F., Addiss, D.G. et al., An outbreak of cryptosporidiosis from fresh-pressed apple cider, *JAMA,* 272, 1592–1596, 1994.
44. Anderson, S., Recent FDA juice HACCP regulations, *Food Safety,* 7, 18–25, 2001.
45. Deng, M.Q. and Cliver, D.O., Inactivation of *Cryptosporidium parvum* oocysts in cider by flash pasteurization, *J. Food Prot.,* 64, 523–527, 2001.
46. Deng, M.Q. and Cliver, D.O., Comparative detection of *Cryptosporidium parvum* oocysts from apple juice, 2000.
47. Centers for Disease Control and Prevention (CDC), An outbreak of Norwalk-like virus associated with a juice processor in Georgia: possible environmental health antecedents, The Environmental Health Service Branch, National Center for Environmental Health, Centers for Disease Control and Prevention, July 5, 2000.
48. Pitt, J.I. and Hocking, A.D., *Fungi and Food Spoilage,* 2nd ed, Blackie Academic and Professional, London, 1997.
49. Frisvad, J.C. and Samson, R.A., Mycotoxin production by food-borne fungi, in *Introduction to Food-borne Fungi,* Samson, R.A., Hoekstra, E.S., Frisvad, J.C., and Filtenborg, O., Eds, Centraalbureau Voor Schimmelcultures, Baarn, The Netherlands, 1996, pp. 251–260.
50. Watkins, K.L., Fazekas, G., and Palmer, M.V., Patulin in Australian apple juice, *Food Aust.,* 42, 438–439, 1990.
51. Yurdun, T., Omurtag, G.Z., and Ersoy, Ö., Incidence of patulin in apple juices marketed in Turkey, *J. Food Prot.,* 64, 1851–1853, 2001.
52. Gökmen, V. and Acar, J., Long-term survey of patulin in apple juice concentrates produced in Turkey, *Food Add. Contam.,* 17, 933–936, 2000.
53. Bracket, R.E., and Marth, E.H., Patulin in apple juice from roadside stands in Wisconsin, *J. Food Prot.,* 42, 862–863, 1979.
54. Marth, E.H., Mycotoxins: production and control, *Food Lab. News,* 8, 34–51, 1992.
55. Nartowicz, V.B., Buchanan, R.L., and Segall, S., Aflatoxin production in regular and decaffeinated coffee beans, *J. Food Sci.,* 56, 1735–1740, 1979.
56. Ochratoxin A in coffee prompts U.K. research effort, *World Food Chem. News,* 1, 14–15, 1995.
57. Parish, M.E., Coliforms, *Escherichia coli* and *Salmonella* serovars associated with a citrus-processing facility implicated in a salmonellosis outbreak, *J. Food Proc.,* 61(3): 280–284, 1998.

58. Tabershaw, I.R, Schmezler, L.L., and Bruyn, H.B., Gastroenteritis from an orange juice preparation I. Clinical and epidemiological aspects, *Arch. Environ. Health,* 15, 72–77, 1967.
59. Schmezler, L.L., Gates, J.M., Redfearn, M.S., and Tabershaw, I.R., Gastroenteritis from an orange juice preparation II. Field and laboratory investigations, *Arch. Environ. Health,* 15, 78–82, 1967.
60. Steele, B.T., Murphy, N., Arbus, G.S., and Rance, C., An outbreak of hemolytic uremic syndrome associated with ingestions of fresh apple juice, *J. Pediatr.,* 101, 963–965, 1982.
61. Centers for Disease Control and Prevention (CDC), Hepatitis A associated with consumption of fresh strawberries — Michigan, *MMWR,* 46, 288–289, 1997.
62. Centers for Disease Control and Prevention (CDC), Update: outbreaks of cyclosporiasis — United States and Canada, *MMWR,* 46, 4–8, 1997.
63. McLellan, M.R. and Splittstoesser, D.F., Reducing risk of *E. coli* in apple cider, *Food Technol.,* 50(12), 174, 1996.
64. D'Aoust, J.Y., Foodborne salmonellosis: current international concerns, *Food Safety Magazine,* 7(2), 10–17, 2000.
65. Singh, B.R., Kulshreshtha, S.B., and Kapoor, K.N., 1995. An orange juice–borne diarrheal outbreak due to enterotoxigenic *Escherichia coli,* I, *Food Sci. Technol.,* 32, 504–506, 1995.
66. Fricker, C. and Smith, H., *Cryptosporidium* and cryptosporidiosis, *SGM Quart.,* 24, 52–53, 1997.
67. Butler, M.A., *Salmonella* outbreak leads to juice recall in Western states, *Food Chem. News,* 42(10), 19–20, 2000.
68. Tamplin, M., *Salmonella* and cantaloupes, *Dairy Food Environ. Sanit.,* 17, 284–286, 1997.
69. Tauxe, R., Kruse, H., Hedberg, C., Potter, M., Madden, J., and Wachsmuth, K., Microbial hazards and emerging issues associated with produce, a preliminary report to the National Advisory Committee on Microbiological Criteria for Foods, *J. Food Prot.,* 60, 1400–1408, 1997.
70. Centers for Disease Control and Prevention (CDC), Multistate outbreak of *Salmonella poona* infections — United States and Canada, *MMWR,* 40, 549–552, 1991.
71. Besser, R.E., Lett, S.M., Weber, J.T., Doyle, M.P., Barrett, T.J., Wells, J.G., and Griffin, P.M., An outbreak of diarrhea and hemolytic uremic syndrome from *Escherichia coli* O157:H7 in fresh-pressed apple cider, *JAMA,* 269, 2217–2220, 1993.
72. Cook, K.A., Dobbs, T.E., Hlady, G., Wells, J.G., Barrett, T.J., Puhr, N.D., Lancette, G.A., Bodager, D.W., Toth, B.L., Genese, C.A., Highsmith, A.K., Pilot, K.E., Finelli, L., and Swerdlow, D.L., Outbreaks of *Salmonella* serotype *hartford* infections associated with unpasteurized orange juice, *JAMA,* 280, 1504–1509, 1998.
73. Whatcom County (Washington) Health Department, A Summary of a Suspected Outbreak of *E. coli* O157:H7 Associated with Consumption of Unpasteurized Apple Cider, 1996, pp. 1–2.

74. Reid, T.M.S. and Robinson, H.G., Frozen raspberries and hepatitis, *Epidemiol. Infect.,* 98, 109–112, 1987.
75. National Advisory Committee on Microbiological Criteria for Foods, NACMCF Recommendations on Fresh Juice, April 1997.
76. Food and Drug Administration (FDA) Center for Food Safety and Applied Nutrition (CFSAN), *Guidance for Industry: Guide to Minimize Microbial Food Safety Hazards for Fresh Fruits and Vegetables*, October 26, 1998, available at http://www.foodsafety.gov/~dms/prodguid.html, retrieved on 1/11/02.
77. U.S. Food and Drug Administration Center for Food Safety and Applied Nutrition, Report of 1997 Inspections of Fresh Unpasteurized Apple Cider Manufacturers: Summary of Results, January 1999, available at http://vm.cfsan.fda.gov/~dms/ciderrpt.html, retrieved on Jan. 28, 2002.
78. U.S. Food and Drug Administration Center for Food Safety and Applied Nutrition, Report of 1997 Inspections of Fresh Unpasteurized Apple Cider Manufacturers: Summary of Results: Analysis of Inspectional Findings, January 1999, available at http://vm.cfsan.fda.gov/~dms/ciderrpt3.html, retrieved on Jan. 28, 2002.
79. U.S. Food and Drug Administration Center for Food Safety and Applied Nutrition, Report of 1997 Inspections of Fresh Unpasteurized Apple Cider Manufacturers: Summary of Results: Attachment 3. Summary of Good Manufacturing Practices Applied by Selected Apple Juice/Cider Producing States and Canada. Analysis of Inspectional Findings, January 1999, available at http://vm.cfsan.fda.gov/~dms/ciderrpt7.html, retrieved on Jan. 28, 2002.
80. National Advisory Committee on Microbiological Criteria for Foods, Hazard Analysis Critical Control Point Principles and Application Guideline, Aug. 14, 1997.
81. Schmidt, R.H., Sims, C.A., Parish, M.E., Pao, S., and Ismail, M., A Model HACCP Plan for Small-scale, Fresh-squeezed (Not Pasteurized) Citrus Juice Operations, Publication CIR 1179, University of Florida, 1997.
82. Senkel, I.A., Jr., Henderson, R.A., Jolbitado, B., and Meng, J., Use of hazard analysis critical control point and alternative treatments in the production of apple cider, *J. Food Prot.,* 62, 778–785, 1999.
83. Kourtis, L.K. and Arvanitoyannis, I.S., Implementation of hazard analysis critical control point (HACCP) system to the nonalcoholic beverage industry, *Food Rev. Int.,* 17, 451–486, 2001.

6 U.S. Food and Drug Administration: Juice HACCP — The Final Rule

Donald A. Kautter, Jr.

CONTENTS

Introduction..125
Concerns with Juice..126
    Microbial Outbreaks 126
    Illnesses from Hazards That Are Not Heat Treatable....................128
    Underreporting..130
    Pesticides..130
    FDA's Public Meeting 132
Consideration of How to Address Juice Concerns 135
    Current Regulation of Juice..135
    The Current Inspection System...135
    Alternatives 136
        Increased Inspection...136
        CGMPs...136
        Mandatory Pasteurization 137
        Labeling..139
        Education...140
        The HACCP Option...140
    Decision to Mandate HACCP 141
    The Final Rule 143
        Pathogen Reduction 151
References...153

INTRODUCTION

The Food and Drug Administration (FDA or the agency) is adopting final regulations to ensure the safe and sanitary processing of fruit and vegetable juices. The regulations mandate the application of Hazard Analysis and Critical Control Point (HACCP) principles to the processing of these foods. HACCP is a preventive system of hazard control. FDA is taking this action because a number of food hazards have been associated with juice products and because a system of preventive control measures is the most effective and efficient way to ensure that these products are safe.

CONCERNS WITH JUICE

Microbial Outbreaks

The Seattle–King County Department of Public Health and the Washington State Department of Health reported, on October 30, 1996, an outbreak of *Escherichia coli* O157:H7 infections epidemiologically associated with drinking a particular brand of unpasteurized apple juice, or juice mixtures containing unpasteurized apple juice, purchased from a coffee shop chain, grocery stores, and other locations (CDC, 1996a). A case was defined as hemolytic uremic syndrome (HUS) or a stool culture yielding *E. coli* O157:H7 in a person who became ill after September 30, 1996, after drinking the particular brand of juice within 10 days before illness onset. At least 66 cases of illness, with 14 cases of HUS and the death of one child, were associated with this outbreak (Griffin, 1996). Cases occurred in British Columbia, California, Colorado, and Washington. *E. coli* O157:H7 isolates cultured from a previously unopened container of the particular brand of apple juice had a deoxyribonucleic acid (DNA) "fingerprint" pattern (restriction fragment length polymorphism) indistinguishable from case-related isolates (CDC, 1996a).

Various juices have been documented as vehicles for causing disease outbreaks from microorganisms. A 1967 outbreak from contaminated water added to orange juice concentrate affected approximately 5,200 persons and was caused by an unidentified virus and possibly other contaminants (Tabershaw et al., 1967; Schmelzer et al., 1967). About 300 people became ill from *Salmonella* serotype *typhimurium* in cider made from apples, including some that had been picked up from the ground in an orchard fertilized with manure, in a 1974 outbreak in New Jersey (CDC, 1975). A 1991 outbreak of *Vibrio cholerae* was associated with coconut milk contaminated during manufacturing in Thailand (CDC, 1991).

There have been two *Cryptosporidium* outbreaks related to drinking apple cider, the first in Maine in 1993 and the other in New York state in 1996. In the first case, the apples used for cider came from trees near a cow pasture (Millard et al., 1994), and in the second case, water used for rinsing came from a well that tested positive for coliforms (CDC, 1996b).

In 1995, an outbreak occurred in Florida that was caused by *Salmonella* serotype *hartford* in unpasteurized orange juice (Cook, 1995). In early 1999 in south Florida, 16 reported cases from *Salmonella* serotype *typhi* were linked to the consumption of frozen mamey, a product often used to make juice beverages (FDA, 1999). During June 1999, there was an outbreak of *Salmonella* serotype *muenchen* infection associated with consumption of unpasteurized orange juice (Anonymous, 1999). As of April 2000, a total of 423 cases, including one that contributed to a death, from *S. muenchen* infection had been reported. Nine additional *Salmonella* serotypes were identified from orange juice collected from the implicated firm.

While no illnesses were reported in October 1998, the state of Florida found *Salmonella manhattan* in an unpasteurized juice blend containing strawberry, apple, and papaya juices (State of Florida, 1998). In November 1999, the same firm involved in the June 1999 outbreak initiated and subsequently expanded a recall because their routine testing found *Salmonella* in samples of unpasteurized orange juice (FDA, 2000). The product had been distributed to restaurants and other food service establishments in eight U.S. states and one Canadian province and to one retail store in Oregon. No known illnesses were associated with this incident.

In April 2000, an outbreak of *Salmonella enteritidis* occurred that was associated with unpasteurized orange juice (Racer, 2000). As of May 2000, 143 cases traced to this orange juice had been identified in Arizona, California, Colorado, Minnesota, Nevada, Washington, and Wyoming. Also in April 2000, 24 people who attended a conference in Atlanta were reported ill with viral gastroenteritis (CDC, 2000). Fresh-squeezed unpasteurized fruit smoothies were implicated in this outbreak. CDC detected Norwalk-like virus in three patient stools.

E. coli O157:H7 has been recognized relatively recently as a human pathogen and has been a source of a number of outbreaks related to juice. Thirteen and possibly 14 children had bloody diarrhea and developed HUS in Toronto between September 15 and 25, 1980. The children's illnesses were associated with drinking fresh apple juice. The children's stools were examined for enteropathogenic *E. coli, Campylobacter, Salmonella, Shigella,* and *Yersinia*. None of these organisms was found. *E. coli* O157:H7 is the suspected causative organism. Conclusive testing for that organism was not performed because *E. coli* O157:H7 was not recognized as a human pathogen before 1982 (Steele, 1982). A 1991 *E. coli* O157:H7 outbreak in southeast Massachusetts conclusively showed that fresh-pressed unpasteurized apple juice can transmit *E. coli* O157:H7 bacteria. In this outbreak, 23 individuals had diarrhea, 16 had bloody diarrhea, and four developed HUS (Besser et al., 1993). In Connecticut, a 1996 outbreak of *E. coli* O157:H7 illness was associated with drinking a particular brand of apple cider. There were 14 cases of illness (including seven hospitalized), with three cases of HUS associated with the outbreak (CDC, 1996b). A small outbreak of *E. coli*

O157:H7 illness in Washington state in 1996 was related to apple cider made at a church event. The apples were washed in a chlorine solution, but it was not reported how much chlorine was used. Six people became ill, but no estimate was given on how many people may have drunk the apple cider (Whatcom County, 1996). In October 1999, there was an outbreak of *E. coli* O157:H7 in commercially processed unpasteurized apple cider in Oklahoma with nine illnesses (seven children) and six hospitalizations (four cases of HUS) (OSDH, 1999).

FDA's recall data and state investigations provide additional evidence of microbial hazards in juice. A 1989 outbreak in New York was caused by the presence in orange juice of *Salmonella* serotype *typhi* that originated from an infected worker and resulted in 69 illnesses with 21 individuals hospitalized (Cambridge, 1997). The state of Washington reported that in 1993 one individual was hospitalized from homemade carrot juice found to contain *Clostridium botulinum* (Walker, 1997). A 1993 Ohio outbreak caused by yeast or some other unknown toxicant in orange juice resulted in 23 illnesses (Karam, 1997). A homemade watermelon drink contaminated with *Salmonella* spp. caused illness in 18 individuals in a 1993 Florida outbreak (Hammond, 1997). The state of Colorado reported two outbreaks of gastrointestinal illness from fresh-squeezed orange juice at a mountain resort (Shillam, 1997).

The evidence shows that certain juices have been the vehicle for outbreaks of foodborne illnesses. Although fruit juice is acidic, and thus would generally be considered to inhibit the growth of most microorganisms, most juice-related outbreaks have been associated with fruit juices.

ILLNESSES FROM HAZARDS THAT ARE NOT HEAT TREATABLE

Illnesses caused by hazards that cannot be reduced to acceptable levels by heat treatment have also been associated with juice. Tin in canned tomato juice caused illness in 113 individuals in 1969 (Barker, 1969). Soil nitrate had resulted in a high nitrate content in the tomatoes, and this high nitrate content accelerated detinning in the cans. In 1984, 11 persons became ill from consuming elderberry juice that contained poisonous parts of the plant; the juice had been prepared by the staff of a religious/philosophic group (CDC, 1984). A 1990 guanabana juice outbreak was caused by the presence of toxic guanabana seed material and caused illness in nine individuals (Hendricks, 1997). A 1997 outbreak was caused by tin in pineapple juice (FDA, 1997a–c).

In 1992, an 18-month-old child with a blood lead level of 36 micrograms per deciliter was found in a routine county health department blood lead monitoring program. Investigation of this incident by the county health department revealed that the only significant source of lead exposure for this child was lead in imported fruit juice packed in 12-ounce, lead-soldered cans (FDA, 1992a–c). Analysis by the state health department of multiple flavors of the fruit juices in lead-soldered cans available to the child found lead levels ranging from 160 to 810 parts per billion (ppb). An exposure assessment performed by the county health department estimated that the child consumed about three cans of these fruit juices per day and estimated that the child's daily lead intake from these fruit juices was approximately 600 μg/day (FDA, 1992a–c). As a result of this incident, FDA announced an emergency action level of 80 ppb for lead in fruit beverages (such as juices, nectars, and drinks) packed in lead-soldered cans (58 FR 17233, April 1, 1993). The agency subsequently banned the use of lead-soldered cans (60 FR 33106, June 27, 1995).

Recalls also provide evidence of non-heat-treatable hazards in juice. In 1988, a fruit punch drink was recalled because of the presence of tin caused by the acidity of the drink reacting with the tin coating of the cans (FDA, 1988a,b). The product had been packaged in the wrong container. There were 10 recalls between 1990 and 1995 for fruit juice or beverages containing fruit juice because of the presence of food ingredients that were inadvertently added to the product, not declared on the label, or not suitable for the food. Food ingredients involved with these recalls were natamycin (FDA, 1991a–c), sulfites (FDA, 1995a–c), FD&C yellow No. 5 (FDA, 1988a,b, 1989, 1990, 1992a–c, 1993a,b), and salt (FDA, 1995a–c). Five recalls between 1991 and 1997 were caused by improper sanitation procedures or faulty equipment. In 1991, sodium hydroxide from a clean-in-place system contaminated the caps of a citrus punch drink (FDA, 1991a–c). In 1992, three persons became ill, with one hospitalized, from a sodium hydroxide sanitizing agent that got into fruit drink product containers during cleaning (FDA, 1992a–c). In 1993, cracks in a heat exchanger allowed an orange-flavored soft drink containing pear juice to come in contact with copper pipe fittings and thus to become contaminated with copper (FDA, 1993a,b). In 1994, milk was found in orange juice from filler lines that were not cleaned between milk and juice production (FDA, 1994a,b). In 1997, the presence of an alkaline cleaning solution in a berry juice caused gastrointestinal distress in several persons (FDA, 1997a–c).

Companies have recalled fruit drinks because pieces of glass or plastic were found in the products. The presence of glass in products is typically caused by the use of glass bottles, which can chip or shatter during the production process (FDA, 1991a–c, 1994a,b, 1997a–c). The plastic was present from the company's practice of draping plastic bags over the side of the bottle-loading bin (FDA, 1996a–c).

One company recalled apple-prune juice and prune juice in 1996 because of unacceptable levels of lead (FDA, 1996a–c). The cause was contaminated imported prune juice.

In response to the establishment of maximum levels for patulin in apple juice by several foreign governments, FDA initiated a sampling survey to determine the levels commonly found in domestic and imported apple juice. Patulin may be present in juice made from moldy apples. In March 1997, the agency found inordinately high levels of patulin in apple juice from a processor in Washington state (Trucksess, 1997). The level of patulin found in the product was sufficient to pose a health hazard, especially considering the fact that apple juice is commonly used by infants and young children (Wagstaff, 1997). All affected products that had left the plant had been used in the manufacture of fermented apple cider. Patulin could not be detected in the fermented product, and it was assumed that the patulin was destroyed through the fermentation process.

Therefore, as the foregoing discussion reveals, the evidence demonstrates that juice and juice beverages are susceptible to chemical and physical hazards as well as microbiological hazards.

Underreporting

There is wide agreement that the laboratory-confirmed cases from outbreaks and recalls understate the actual number of juice-related cases, but no consensus exists on the extent of the understatement. Individuals may not manifest all symptoms or have severe enough symptoms to necessitate medical attention. Medical personnel may simply treat an individual's symptoms without determining the underlying cause. The laboratory-confirmed cases only represent those cases where individuals sought medical attention and where medical personnel performed the necessary testing and reported the case to a government agency. While the actual number of juice-related illnesses is unknown, FDA has derived an estimate of the total number by multiplying the average number of laboratory-confirmed cases by factors that account for underreporting. The factors are based on the relationships between annual outbreak cases and published estimates of the number of foodborne illnesses. For example, using these adjustment factors, it is estimated that the average 16 annual laboratory-confirmed cases of *Salmonella* represent 4900 to 7600 actual cases (Williams et al., 1997). For *E. coli* O157:H7, an average 22 laboratory-confirmed cases per year may actually represent 2200 to 4300 total juice-related cases (Williams et al., 1997). Therefore, the agency assumes that the actual number of illnesses from the outbreaks described in the previous sections of this document is much greater than the confirmed number of illnesses.

Pesticides

Pesticides are usually applied to plants to combat insects, plant diseases, and weed growth to assist in the growth of the fruit or vegetable. A food is considered adulterated under Section 402(a)(2)(B) of the Federal Food, Drug and Cosmetic Act (the act) (21 U.S.C. 342(a)(2)(B)) if pesticide residues are present above the Environmental Protection Agency (EPA) established tolerances, or if EPA has not established a tolerance for use of the pesticide on the particular plant.

FDA annually monitors a wide variety of foods for pesticide residues. In 1994, FDA sampled 1411 domestic fruits and fruit products, including apple juice and other fruit juices, for pesticide residues and found that less than 1 percent were violative for being over tolerance and less than 1 percent were violative for having no tolerance (FDA, 1995a–c). None of the 122 samples of apple juice or 44 samples of other fruit juices were violative.

Out of 1795 samples of domestic vegetables and vegetable products tested, FDA found that less than 1 percent of samples were over tolerance, and 2 percent were violative for having no tolerance. FDA also tested 1940 imported fruits and fruit products in its 1994 pesticide residue–monitoring program. Less than 1 percent of the items tested were over tolerance and 3 percent were violative for having no tolerance. None of the 110 fruit juices sampled were violative. The agency sampled 2460 imported vegetables and vegetable products and found that less than 1 percent were violative for being over tolerance and 4 percent for having no tolerance.

In its 1995 pesticide monitoring program, FDA found less than 1 percent of 1437 samples of domestic fruits and fruit products to be violative for being over tolerance and 1 percent to be violative for having no tolerance (FDA, 1996a–c). Of the 110 apple juices and 22 other fruit juices sampled, only a single apple juice sample was found to be violative because of the presence of a pesticide with no established tolerance. Analysis of 1585 samples of domestic vegetables and vegetable products produced results similar to the results found in 1994, i.e., less than 1 percent of samples were over tolerance, and approximately 2 percent were violative because there were no tolerances for the pesticide residues that FDA found.

The agency sampled 1757 imported fruits and fruit products for pesticides in 1995 and found that less than 1 percent were violative for being over tolerance and that 3 percent were violative for having no tolerance. Of the 19 apple juices and 52 other fruit juices tested, two apple juice samples were violative because they contained pesticides for which there were no established tolerances. The agency sampled 2535 imported vegetables and vegetable products and found that 1 percent were violative for being over tolerance and that 3 percent were violative for having pesticide residues for which there was no tolerance. Some of these samples contained both residues over tolerance and residues with no tolerance.

Although there are no documented outbreaks of illness caused by unlawful pesticide residues, chronic exposure to pesticide residues that do not conform to EPA tolerances increases risks to the public health. Therefore, juice processors must determine whether the possible presence of unlawful pesticide residues is a hazard that is reasonably likely to occur.

FDA's Public Meeting

As a result of the October 1996 apple juice outbreak from *E. coli* O157:H7, FDA held a public meeting on December 16 and 17, 1996 (hereafter referred to as the juice meeting), to review the current science, including technological and safety factors, relating to fresh juices and to consider measures necessary to provide safe fruit juices to the public. Interested persons were given until January 3, 1997, to submit written comments on the notice. On January 2, 1997 (62 FR 102), FDA extended the comment period to February 3, 1997, in response to several requests for an extension. The purpose of the juice meeting was to provide a forum for an information exchange on current industry practices for the production of juice products and on developments in the science underlying the production of safe juices. Experts from industry, academia, and the regulatory and consumer sectors presented information on illnesses and the epidemiology of outbreaks arising from contaminated juices; concerns about emerging pathogens; the *E. coli* O157:H7 outbreak in October 1996 caused by contaminated apple juice; procedures for processing juices; and new and existing technology to remove or decrease the number of pathogens or other contaminating microorganisms. The meeting provided an opportunity to:

1. Consider how FDA's regulatory program for fresh juice and juice products should be revised
2. Discuss and exchange information on relevant safety issues
3. Identify research needs where appropriate
4. Consider whether additional consumer education is necessary
5. Consider whether other measures were needed to reduce the risk of future outbreaks of illness from juice

FDA received over 180 comments from industry (with a number of these describing themselves as small businesses), consumers, consumer organizations, trade organizations, scientific/technical companies, academic institutions or organizations, state agencies, a local government agency, and members of Congress. Although most of the comments concerned apple juice specifically, many comments pertained to juices in general, and some referred only to citrus juices. Most comments were concerned with changes in processing to improve the safety of juices. Among the changes suggested were requiring pasteurization of juices, requiring HACCP, or establishing current good manufacturing practices (CGMPs) in juice processing. The agency addressed the comments made at the meeting or submitted in response to the Federal Register notice in the juice HACCP proposal. The Fresh Produce Subcommittee (FPS) of the National Advisory Committee on Microbiological Criteria for Foods (NACMCF) attended the public meeting. The FPS met after the public meeting and made recommendations to the NACMCF. The NACMCF subsequently met to discuss the issues that were raised at the meeting. Based on information that was presented at the meeting and on the FPS's expertise, the full NACMCF made several recommendations (NACMCF, 1997).

The NACMCF stated that there are many aspects that affect pathogen control, such as:

- Agricultural practices
- Product handling
- Equipment used
- Growing location, including produce obtained from below ground (carrots), on ground (e.g., tree drops), or picked from trees
- pH
- Acidulants
- Method of processing
- Degree of animal contact
- Refrigeration
- Packaging
- The distribution system

It stated that, in determining the best control mechanisms, it is important to remember that the conditions for microbial survival differ from those for growth. The NACMCF recognized that while the risks associated with specific juices vary, there are safety concerns associated with juices, especially unpasteurized juices. The NACMCF concluded:

1. The history of public health problems associated with fresh juices indicates a need for active safety interventions.
2. For some fruit (e.g., oranges), the need for intervention may be limited to surface treatment, but for others, additional interventions may be required (e.g., pasteurization of the juice).

The NACMCF recommended to FDA the use of safety performance criteria instead of mandating the use of a specific intervention technology. In the absence of known specific pathogen–product associations, the NACMCF recommended the use of *E. coli* O157:H7 or *Listeria monocytogenes* as the target organism, as appropriate. This recommendation was based on the premise that these organisms are two of the most difficult to control (i.e., by juice acidity or heat lethality), and that, by controlling them, other pathogenic organisms will likely be controlled. The NACMCF suggested that a tolerable level of risk may be achieved by requiring interventions that have been validated to achieve a cumulative 5-log reduction in the target pathogen or a reduction in yearly risk of illness to less than 10^{-5}, assuming consumption of 100 ml of juice daily. In addition, the NACMCF stated that HACCP and safety performance criteria should form the general conceptual framework to ensure the safety of juices, and that control measures should be based on a thorough hazard analysis.

The NACMCF also stated that validation of the process must be an integral part of this framework. The NACMCF recommended mandatory HACCP for all juice products, and that processors should implement and strictly adhere to industry CGMPs. The NACMCF also recommended industry education programs addressing basic food microbiology, the principles of cleaning and sanitizing equipment, CGMPs, and HACCP. The NACMCF recommended further study in several areas:

1. The efficacy of new technologies and intervention strategies for safety
2. The contamination, survival, and growth of pathogens on produce with or without breaks in skin, with or without areas of rot, and within the core
3. How produce becomes contaminated with human pathogens, including the relevant microbial ecology during production and processing of juice (In particular, the NACMCF stated that there is an urgent need for these types of studies on *E. coli* O157:H7 in apple juice.)
4. The baseline incidence of human pathogens on fruits and vegetables, particularly on those used in juice processing
5. Labeling information needed for consumer understanding and choice of safer juices and juice products

On the basis of all the testimony presented at the December 16 and 17, 1996 meeting, the NACMCF agreed that there is a need to understand the differences among juices and juice products (e.g., citrus versus other). A significant problem identified by the NACMCF is that consumers presently do not have a means to clearly differentiate between unpasteurized and pasteurized products, and that terms used to refer to juice products do not always have universal meanings. For example, "cider" is perceived to refer to an unpasteurized product whereas products referred to by the term "juice" are often perceived to be pasteurized. The NACMCF also stated that traditional heat treatments given to juices and juice products have been designed to achieve shelf stability, to remove water (i.e., concentration), or to affect other quality-related factors, and that these treatments, commonly referred to as "pasteurization," are greatly in excess of a process needed to inactivate foodborne pathogens. Because of the lack of sufficient data to evaluate the effectiveness of labeling statements as safety interventions or to inform consumer choice, the NACMCF stated that it could not strongly endorse labeling as an interim safety measure. Although the NACMCF did not endorse labeling as an interim safety measure, the FDA did mandate an interim labeling measure for packaged juice.

CONSIDERATION OF HOW TO ADDRESS JUICE CONCERNS

CURRENT REGULATION OF JUICE

FDA has established labeling regulations and standards of identity for a number of juices. 21 CFR 101.30 pertains to percentage juice declaration for beverages that contain fruit or vegetable juice. Common or usual name regulations for nonstandardized beverages that contain fruit or vegetable juice are found in 21 CFR 102.33. Standards of identity are found in part 146 (21 CFR part 146) for a number of fruit juices and beverages and in part 156 (21 CFR part 156) for tomato juice. The standard of identity for pasteurized orange juice states, "The orange juice is so treated by heat as to reduce substantially the enzymatic activity and the number of viable microorganisms." Pasteurized orange juice must be labeled as such.

THE CURRENT INSPECTION SYSTEM

Juice processors, like other food processors, are subject to periodic unannounced, mandatory inspection by FDA. This inspection system provides the agency with a picture of conditions at a facility at the time of the inspection. However, assumptions must be made about conditions at the facility before and after that inspection, as well as about important factors beyond the facility that have a bearing on the safety of the finished product. The reliability of these assumptions over the intervals between inspections can create questions about the adequacy of the system. FDA's inspections are based, in part, upon its regulations on CGMP in the manufacturing, packing, or holding of human food in part 110 (21 CFR part 110). For the most part, these regulations set out broad statements of general applicability to all food processing on matters such as sanitation, facilities, equipment and utensils, processes, and controls. HACCP-type controls are listed as one of several options available to prevent food contamination (Sec. 110.80(b)(13)(i)), but they are not integral to the controls outlined in the regulations. The inspection and surveillance strategies that FDA uses ascertain a manufacturer's knowledge of hazards and preventive control measures largely by inference (i.e., based on whether a company's products are in fact adulterated, or whether conditions in a plant are consistent with CGMP). It is the manufacturer's responsibility to ensure that its products are in compliance with the act. However, in the face of new pathogens, such as *E. coli* O157:H7, and the risk of illness associated with these pathogens, especially for children, the elderly, and the immunocompromised, FDA concludes that, at least for juices, new measures to control microbial, chemical, and physical hazards are necessary to ensure that finished products comply with the act's standards.

Alternatives

Comments from the juice meeting suggested several alternatives to ensure that juice products are safe, including the following:

Increased Inspection

Continuous visual inspection of juice production is not a viable alternative because few hazards associated with juice are detectable through visual inspection. Another possibility is to direct significant additional resources toward increasing the frequency of FDA's inspection of juice manufacturers, as well as increasing the agency's sampling, laboratory analysis, and related regulatory activities with respect to these products. While many samples of domestic and imported juice products are collected each year for analysis in FDA laboratories, and this sampling is designed to represent a broad range of products and to target known problems, the product sampled represents only a small fraction of the total poundage of juice products consumed in this country. Substantially more expenditures would be needed to increase laboratory analyses to statistically significant levels. Even if the funds for increased FDA inspection and increased sampling and analysis were available, this approach alone would not likely be the best way for the agency to spend its limited resources to protect the public health. Reliance on end product testing involves a certain amount of inefficiency and enormous sample sizes, and testing on a lot-by-lot basis is necessary to overcome that inefficiency. Therefore, this option has significant limitations.

CGMPs

Many comments from the juice meeting urged the implementation of industry CGMPs or sanitation standards to increase the safety of juices. Some comments provided state rules, model CGMPs, or sanitation guidelines for

FDA's consideration. Other comments stated that there is a need for more industry education on sanitation and hygiene. CGMP regulations have a twofold purpose:

1. To provide guidance on how to reduce unsanitary manufacturing practices and on how to protect against food becoming contaminated
2. To set out objective requirements that enable industry to know what FDA expects an investigator to find when he or she visits a food plant (51 FR 22458 at 22459, June 19, 1986)

CGMPs consist generally of broad statements on sanitation, facilities, equipment and utensils, processes, and controls that are of general applicability to food processing. Therefore, FDA issuance of CGMPs for juice would be an approach that could assist manufacturers in the production of safe juices. FDA encourages the juice industry to use CGMPs to help ensure the safety of their juices. As stated previously, the NACMCF recommended that processors implement and strictly adhere to industry CGMPs.

However, the use of CGMPs alone may not be adequate to ensure that juices are safe because of the broad-based nature of CGMPs. CGMPs are directed at plant-wide operating procedures and do not concentrate on the identification and prevention of food hazards. Therefore, the agency concludes that CGMPs, although useful, will not be adequate, without additional measures, to ensure the safety of juices.

Mandatory Pasteurization

The majority of the comments from the juice public meeting pertained to pasteurization of juice. A number of comments urged FDA to mandate pasteurization or other equivalent treatment of fruit juice to ensure its safety. One person who commented reported that consumers of his apple cider had not complained about a difference in flavor when he implemented pasteurization. Another suggested that pasteurization be required for apple cider only if CGMPs and HACCP fail. One comment suggested that pasteurization be required only for apple juice because of the difficulty in cleaning apples as compared to other fruits. However, most comments opposed mandatory pasteurization of juices because of:

1. The expense of pasteurization equipment
2. Preference by some consumers for the flavor of unpasteurized over pasteurized juice
3. The safety record of juices
4. Degradation of nutritional value from heat treatment

Many comments from small businesses claimed that they would be forced to close their operations if pasteurization were required. Some comments also stated an economic need for the use of dropped apples ("drops"), with one recommending the use of only hand-picked (rather than machine-picked) drops. Other comments stated that the use of drops should be prohibited, at least in unpasteurized juices. FDA is aware of the significant safety advantages of pasteurizing juice as well as of the reasons that some processors choose not to pasteurize their products.

Pasteurization is a heat treatment used to inactivate the vegetative forms of specific bacteria in liquid or semi-liquid food products. Pasteurization is an effective and proven technology to ensure that juice does not contain pathogens. However, there may be other methods that are equally effective. Thus, the NACMCF recommended the establishment of safety performance criteria for appropriate target organisms rather than the establishment of a specific intervention technology. The NACMCF stated that safety performance criteria would be most effective.

For example, whole oranges with an intact skin may be processed so that pathogens on the surface of the fruit are destroyed. Because pathogens are not reasonably likely to be present in the interior of an orange, surface treatment could be adequate to ensure the safety of the juice.

This example illustrates that if FDA were to mandate pasteurization, such action could have the effect of limiting the development of new technologies that are as effective as pasteurization in particular circumstances but less intrusive and less expensive. Therefore, the agency concludes that relying on safety performance criteria, as recommended by the NACMCF, is an approach preferable to mandating pasteurization. However, if the use of safety performance criteria does not significantly decrease the number of microbial outbreaks caused by juice, the agency may consider adopting a regulation that mandates pasteurization. The agency disagrees with the comments that stated that it should require that apple juice be pasteurized because apples can be difficult to clean. FDA recognizes that pasteurization is a process that has been validated to meet NACMCF's recommendations. Manufacturers may be able to use other technologies and practices provided that their process is validated to achieve a 5-log reduction in the target pathogen. Therefore, reliance on safety performance criteria is a better long-term approach because it provides for the development of new technologies.

A number of comments at the juice meeting urged FDA to consider alternatives to pasteurization to increase the safety of juices. Alternatives suggested by the comments included extreme isostatic pressure, high pressure sterilization, ultra short time–heat exchanger processing, ohmic heating, aseptic packaging, modified atmosphere packaging, ultrafiltration, high temperature and high pH adjustment of wash water, ultrahigh hydrostatic pressure, electric pulses, electromagnetic field, pulsed light, ultraviolet (UV) water treatment, UV treatment with photoreactivation, electron beam sterilization, irradiation, ozonated water treatment, microbiocidal additives (benzoate, sorbate), and pH adjustment. The comments recommended that sanitizers or ingredients for washes include use of chlorine, chlorous acid, chlorine with emulsifiers, trisodium phosphate, peroxyacetic acid, peracetic acid, or dimethyl dicarbonate.

The agency agrees that there may be a number of agents that can reduce the number of microorganisms present in juice. As the NACMCF recommended, a tolerable level of risk may be achieved by interventions that have been validated to achieve a cumulative 5-log reduction in the target pathogens or a reduction in yearly risk of illness to less than 10^{-5}, assuming consumption of 100 ml of juice daily. However, the NACMCF did not specify the manner in which this risk reduction should be accomplished, only the target that must be reached.

Labeling

A number of comments suggested that labeling to distinguish pasteurized from unpasteurized juice would enable consumers to make an informed choice. One of the comments requested warnings to those "at risk," one urged the publication of warnings in the newspaper, and another wanted labeling with no warning. Rather than labeling, one comment suggested point of sale information. One comment urged FDA not to require labeling to distinguish pasteurized from unpasteurized juices. The NACMCF recommended research on labeling information needed for consumer understanding and choice of safer juice products. The NACMCF concluded that, while the risks associated with specific juices vary, there are safety concerns associated with juices generally, especially unpasteurized juices. Labeling whether a product is pasteurized or unpasteurized is useful information, and the agency encourages processors to place such information on labels. However, such labeling would not inform purchasers of unpasteurized products that children, the elderly, and the immunocompromised are "at risk" from consuming the product. Without effective consumer education, the label statements "pasteurized" and "unpasteurized" are likely to have relatively little meaning to consumers and could even cause confusion because some consumers might select unpasteurized juice, considering it more "healthy" because it is less processed. Finally, a labeling requirement that focuses only on whether a product is pasteurized or unpasteurized does not take into account technologies other than pasteurization that are adequate to control pathogens, and, thus, such a requirement could be viewed as restricting the development of new technologies.

The agency outlined measures in a final rule published July 8, 1998 (63 FR 37030) on labeling for packaged juice. These labeling measures attempt to provide information on the risks that juice that has not been processed to control pathogens poses to children, the elderly, and the immunocompromised.

Education

Other comments from the juice meeting suggested that education would increase the awareness associated with the safety of juices and of all foods. Some comments suggested that more industry education or training was needed. Other comments wanted more consumer education, especially for those at highest risk from foodborne disease. The NACMCF recommended that the industry be educated on basic food microbiology, the principles of cleaning and sanitizing equipment, CGMPs, and HACCP. FDA agrees that industry education can serve a valuable role in controlling potential food hazards and encourages the industry to take an active part in educating its employees and utilizing up-to-date technologies. The agency will assist the industry in its education effort. Concerning consumer education, the agency has launched several initiatives to inform consumers about the potential hazards presented by juice to at-risk individuals (see 62 FR 45593, August 28, 1997). However, no matter how extensive a consumer education initiative the agency undertakes, it is doubtful that consumer education will reach all at-risk consumers. Therefore, consumer education alone will not be adequate to inform the at-risk population of the potential hazards of consumption of juice that has not been processed to control pathogens. Given that effective processing methods are available, primary reliance needs to be placed on them to ensure the safety of juice.

The HACCP Option

Many of the attendees at the juice meeting urged FDA to mandate HACCP for juice processors, whereas others were opposed. A number of the attendees urged use of CGMPs together with HACCP. Some attendees at the juice meeting recommended that microbiological criteria or performance standards be used in addition to HACCP, with two suggesting a 5-log reduction for *E. coli* O157:H7. The NACMCF concluded that HACCP and safety performance criteria can provide the general conceptual framework needed to ensure the safety of juices, and that validation of the HACCP plan for the juice process (i.e., ensuring that the process is adequate to control hazards) must be an integral part of this framework. The NACMCF stated that processors should establish HACCP control measures based on a thorough hazard analysis.

HACCP is a preventive system of hazard control that places the responsibility for identifying safety problems with the manufacturer. Use of the HACCP system means that a firm is engaged in continuous problem prevention and problem solving, rather than relying on facility inspections by regulatory agencies or consumer complaints to detect a loss of control. HACCP provides for real-time monitoring to assess the effectiveness of control. A HACCP system put in place by a manufacturer for a particular facility is unique and must reflect the type of juice, its method of processing, its packaging, the facility in which it is prepared, and the intended consumers. As discussed previously, there is sufficient evidence to demonstrate that significant problems exist with the presence of pathogens in some juice products. Pathogens in juice can be controlled by heat treatment. However, there may be other treatments that meet the same performance standard that are equally effective (e.g., multiple barriers, surface treatment of intact fruit). The use of a HACCP system provides flexibility to a processor to use alternative pathogen control methods and, thus, encourages the development of new technologies but does not dictate either their development or use. Moreover, not only is HACCP effective in controlling microbiological hazards, it also is effective in preventing chemical and physical hazards. Thus, HACCP is particularly well suited for the juice industry given, as discussed previously, the range of hazards that must be addressed in processing juice.

The agency agrees with the comments that urged use of CGMPs together with HACCP. CGMPs form the foundation upon which a HACCP system is built. Therefore, CGMPs are integral to the HACCP approach. Because there are significant concerns with the microbial safety of juices, HACCP systems must control pathogens. FDA is mandating a 5-log reduction in target pathogens, as the NACMCF recommended, as a necessary step in a HACCP plan for juice. Validation of a HACCP system must ensure that the process that is employed is adequate to control the relevant pathogens, in addition to chemical and physical hazards. Validation of performance standards consists of determining the ability of the pathogens in question to resist acid and other chemical or heat treatment and the ability of the process applied to overcome that resistance.

DECISION TO MANDATE HACCP

As discussed above, in developing a strategy to address the hazards associated with juice, FDA considered the following alternatives to HACCP:

1. Increased inspections
2. Current good manufacturing practices (CGMPs)
3. Mandatory pasteurization
4. Labeling as a long-term solution

5. Education
6. An approach that would draw a distinction between untreated apple cider and all other juices

The agency discussed each alternative in the HACCP proposed rule (63 FR 20450 at 20454) and its reasons for mandating the use of HACCP systems rather than the alternatives (FDA, 2001). HACCP is a focused, efficient, preventive system that minimizes the chance that foods contaminated with hazardous materials or microorganisms will be consumed. The strength of HACCP lies in its ability to enable the processor to identify, systematically and scientifically, the primary food safety hazards of concern for the specific products, the specific processes, and the specific manufacturing facilities in question, and then to implement on a focused, consistent basis, steps (critical control points [CCPs]) in food production, processing, or preparation that are critical to prevent, reduce to acceptable levels, or eliminate hazards from the particular food being processed.

Flexibility in how to address identified hazards is inherent in HACCP systems. Even when producing comparable products, no two processors use the same source of incoming materials or the same processing technique, or manufacture in identical facilities. Each of these factors (and their many combinations) presents potential opportunities for contamination of the food. HACCP focuses the processor on understanding his/her own process and the hazards that may be introduced during that process, and identifying specific controls to prevent, reduce, or eliminate the identified hazards. The flexibility of the HACCP approach is a critically important attribute. This flexibility allows manufacturers to adjust CCPs, adjust techniques used to address CCPs when changes occur in the system (e.g., use of new ingredients), and readily incorporate new scientific developments (e.g., use of new control techniques, new preventive technologies, identification of new hazards).

Another important strength of HACCP is the development of a plan written by the processor detailing the control measures to be used at CCPs. By developing a written plan, juice processors gain a working knowledge of their processing system, its effect on the food, and where in the system potential contamination may occur. Both the processor and the agency are able to derive the full benefits of a HACCP system. The hazard analysis and HACCP plan allow both the processor and the agency to verify and validate the operation of the system.

HACCP's flexibility also permits processors to select the appropriate control measures in the context of how the whole system functions, allowing processors to use the most appropriate and economical methods to control food hazards that are reasonably likely to occur in their operations. The ability to choose among various control methods encourages research on and development of new and innovative technologies to better address individual situations. Because of its flexibility, HACCP is particularly advantageous to small businesses and seasonal processors.

HACCP provides the processor with a record of identified food hazards. It allows quick identification of a breakdown in the processing system and thus, prevents products with food hazards from entering the marketplace and causing illness. Moreover, review of records over a longer period of time (days or weeks) may reveal a trend toward a breakdown in the system, such as a critical processing temperature that is slowly drifting down. HACCP records allow evaluation of whether changes in the processing system require changes in CCPs or their critical limits (CLs), thus ensuring that the HACCP system is up to date and adequate to control all food hazards that are reasonably likely to occur. This recordkeeping also allows regulatory investigators to readily review the long-term performance of a firm's processing system, rather than relying on a time-limited inspection, which provides only a snapshot of how well the firm is doing in producing and distributing safe product on any given day.

HACCP is ideally suited to respond to emerging problems because a HACCP system is a dynamic system that must be validated periodically to ensure that all hazards reasonably likely to occur are identified and controlled via CCPs. Validation of both the hazard analysis and the HACCP plan entails a thorough review to ensure that all hazards that are reasonably likely to occur are addressed in the HACCP system.

Because of its preventive yet flexible nature, HACCP is recognized by food safety professionals as the single most effective means to assure the safety of foods. It has been endorsed by the National Academy of Sciences (NAS, 1985), the Codex Alimentarius Commission (an international food standard-setting organization) (Codex, 1997), and the NACMCF (NACMCF, 1998). Increasingly, use of HACCP systems is an indication to importing countries that food safety systems that provide a standardized level of public health protection are in place and being used by producers in exporting countries.

THE FINAL RULE

SECTION 120.1 APPLICABILITY

(a) Any juice sold as such or used as an ingredient in beverages shall be processed in accordance with the requirements of this part. Juice means the aqueous liquid expressed or extracted from one or more fruits or vegetables, purees of the edible portions of one or more fruits or vegetables, or any concentrates of such liquid or puree. The requirements of this part shall apply to any juice regardless of whether the juice, or any of its ingredients, is or has been shipped in interstate commerce (as defined in Section 201(b) of the Federal Food, Drug, and Cosmetic Act, 21 U.S.C. 321(b)). Raw agricultural ingredients of juice are not subject to the requirements of this part. Processors should apply existing agency guidance to minimize microbial food safety hazards for fresh fruits and vegetables in handling raw agricultural products. (b) The regulations in this part shall be effective January 22, 2002. However, by its terms, this part is not binding on small and very small businesses until the dates listed in paragraphs (b)(1) and (b)(2) of this section. (1) For small businesses employing fewer than 500 persons the regulations in this part are binding on January 21, 2003. (2) For very small businesses that have either total annual sales of less than $500,000, or if their total annual sales are greater than $500,000 but their total food sales are less than $50,000; or the person claiming this exemption employed fewer than an average of 100 full-time equivalent employees and fewer than 100,000 units of juice were sold in the United States, the regulations are binding on January 20, 2004.

SECTION 120.3 DEFINITIONS

The definitions of terms in Section 201 of the Federal Food, Drug, and Cosmetic Act, Sec. 101.9(j)(18)(vi), and part 110 of this chapter are applicable to such terms when used in this part, except where redefined in this part. The following definitions shall also apply: (a) Cleaned means washed with water of adequate sanitary quality. (b) Control means to prevent, eliminate, or reduce. (c) Control measure means any action or activity to prevent, reduce to acceptable levels, or eliminate a hazard. (d) Critical control point means a point, step, or procedure in a food process at which a control measure can be applied and at which control is essential to reduce an identified food hazard to an acceptable level. (e) Critical limit means the maximum or minimum value to which a physical, biological, or chemical parameter must be controlled at a critical control point to prevent, eliminate, or reduce to an acceptable level the occurrence of the identified food hazard. (f) Culled means separation of damaged fruit from undamaged fruit. For processors of citrus juices using treatments to fruit surfaces to comply with Section 120.24, culled means undamaged, tree-picked fruit that is U.S. Department of Agriculture choice or higher quality. (g) Food hazard means any biological, chemical, or physical agent that is reasonably likely to cause illness or injury in the absence of its control. (h) Importer means either the U.S. owner or consignee at the time of entry of a food product into the United States, or the U.S. agent or representative of the foreign owner or consignee at the time of entry into the United States. The importer is responsible for ensuring that goods being offered for entry into the United States are in compliance with all applicable laws. For the purposes of this definition, the importer is ordinarily not the custom house broker, the freight forwarder, the carrier, or the steamship representative. (i) Monitor means to conduct a planned sequence of observations or measurements to assess whether a process, point, or procedure is under control and to produce an accurate record for use in verification. (j)(1) Processing means activities that are directly related to the production of juice products. (2) For purposes of this part, processing does not include: (i) Harvesting, picking, or transporting raw agricultural ingredients of juice products, without otherwise engaging in processing; and (ii) The operation of a retail establishment. (k) Processor means any person engaged in commercial, custom, or institutional processing of juice products, either in the United States or in a foreign country, including any person engaged in the processing of juice products that are intended for use in market or consumer tests. (l) Retail establishment is an operation that provides juice directly to the consumers and does not include an establishment that sells or distributes juice to other business entities as well as directly to consumers. "Provides" includes storing, preparing, packaging, serving, and vending. (m) Shall is used to state mandatory requirements. (n)

U.S. Food and Drug Administration: Juice HACCP — The Final Rule

Shelf-stable product means a product that is hermetically sealed and, when stored at room temperature, should not demonstrate any microbial growth. (o) Should is used to state recommended or advisory procedures or to identify recommended equipment. (p) Validation means that element of verification focused on collecting and evaluating scientific and technical information to determine whether the HACCP plan, when properly implemented, will effectively control the identified food hazards. (q) Verification means those activities, other than monitoring, that establish the validity of the HACCP plan and that the system is operating according to the plan.

SECTION 120.5 CURRENT GOOD MANUFACTURING PRACTICE

Part 110 of this chapter applies in determining whether the facilities, methods, practices, and controls used to process juice are safe, and whether the food has been processed under sanitary conditions.

SECTION 120.6 SANITATION STANDARD OPERATING PROCEDURES

(a) Sanitation controls. Each processor shall have and implement a sanitation standard operating procedure (SSOP) that addresses sanitation conditions and practices before, during, and after processing. The SSOP shall address: (1) Safety of the water that comes into contact with food or food contact surfaces or that is used in the manufacture of ice; (2) Condition and cleanliness of food contact surfaces, including utensils, gloves, and outer garments; (3) Prevention of cross contamination from insanitary objects to food, food packaging material, and other food contact surfaces, including utensils, gloves, and outer garments, and from raw product to processed product; (4) Maintenance of handwashing, hand sanitizing, and toilet facilities; (5) Protection of food, food packaging material, and food contact surfaces from adulteration with lubricants, fuel, pesticides, cleaning compounds, sanitizing agents, condensate, and other chemical, physical, and biological contaminants; (6) Proper labeling, storage, and use of toxic compounds; (7) Control of employee health conditions that could result in the microbiological contamination of food, food packaging materials, and food contact surfaces; and (8) Exclusion of pests from the food plant. (b) Monitoring. The processor shall monitor the conditions and practices during processing with sufficient frequency to ensure, at a minimum, conformance with those conditions and practices specified in part 110 of the Food, Drug and Cosmetic Act that are appropriate both to the plant and to the food being processed. Each processor shall correct, in a timely manner, those conditions and practices that are not met. (c) Records. Each processor shall maintain SSOP records that, at a minimum, document the monitoring and corrections prescribed by paragraph (b) of this section. These records are subject to the recordkeeping requirements of Section 120.12. (d) Relationship to Hazard Analysis and Critical Control Point (HACCP) plan. Sanitation standard operating procedure controls may be included in the HACCP plan required under Section 120.8(b). However, to the extent that they are implemented in accordance with this section, they need not be included in the HACCP plan.

SECTION 120.7 HAZARD ANALYSIS

(a) Each processor shall develop, or have developed for it, a written hazard analysis to determine whether there are food hazards that are reasonably likely to occur for each type of juice processed by that processor and to identify control measures that the processor can apply to control those hazards. The written hazard analysis shall consist of at least the following: (1) Identification of food hazards; (2) An evaluation of each food hazard identified to determine if the hazard is reasonably likely to occur and thus, constitutes a food hazard that must be addressed in the HACCP plan. A food hazard that is reasonably likely to occur is one for which a prudent processor would establish controls because experience, illness data, scientific reports, or other information provide a basis to conclude that there is a reasonable possibility that, in the absence of those controls, the food hazard will occur in the particular type of product being processed. This evaluation shall include an assessment of the severity of the illness or injury if the food hazard occurs; (3) Identification of the control measures that the processor can apply to control the food hazards identified as reasonably likely to occur in paragraph (a)(2) of this section; (4) Review of the current process to determine whether modifications are necessary; and (5) Identification of critical control points. (b) The hazard analysis shall include food hazards that can be introduced both within and outside the processing plant environment, including food hazards that can occur before, during, and after harvest. The hazard analysis shall be developed by an individual or individuals who have been trained in accordance with Section 120.13 and shall be subject to the recordkeeping requirements of Sec. 120.12. (c) In evaluating what food hazards are reasonably likely to occur, consideration should be given, at a minimum, to the following: (1) Microbiological contamination; (2) Parasites; (3) Chemical contamination; (4) Unlawful pesticide residues; (5) Decomposition in food where a food hazard has been associated with decomposition; (6) Natural toxins; (7) Unapproved use of food or color additives; (8) Presence of undeclared ingredients that may be allergens; and (9) Physical hazards. (d) Processors should evaluate product ingredients, processing procedures, packaging, storage, and intended use; facility and equipment function and design; and plant sanitation, including employee hygiene, to determine the potential effect of each on the safety of the finished food for the intended consumer. (e) HACCP plans for juice need not address the food hazards associated with microorganisms and microbial toxins that are controlled by the requirements of part 113 or part 114 of this chapter. A HACCP plan for such juice shall address any other food hazards that are reasonably likely to occur.

SECTION 120.8 HAZARD ANALYSIS AND CRITICAL CONTROL POINT (HACCP) PLAN

(a) HACCP plan. Each processor shall have and implement a written HACCP plan whenever a hazard analysis reveals one or more food hazards that are reasonably likely to occur during processing, as described in Section 120.7. The HACCP plan shall be developed by an individual or individuals who have been trained in accordance with Sec. 120.13 and shall be subject to the recordkeeping requirements of Sec. 120.12. A HACCP plan shall be specific to: (1) Each location where juice is processed by that processor; and (2) Each type of juice processed by the processor. The plan may group types of juice products together, or group types of production methods together, if the food hazards, critical control points, critical limits, and procedures required to be identified and performed by paragraph (b) of this section are essentially identical, provided that any required features of the plan that are unique to a specific product or method are clearly delineated in the plan and are observed in practice. (b) The contents of the HACCP plan. The HACCP plan shall, at a minimum: (1) List all food hazards that are reasonably likely to occur as identified in accordance with Sec. 120.7, and that thus must be controlled for each type of product; (2) List the critical control points for each of the identified food hazards that is reasonably likely to occur, including as appropriate: (i) Critical control points designed to control food hazards that are reasonably likely to occur and could be introduced inside the processing plant environment; and (ii) Critical control points designed to control food hazards introduced outside the processing plant environment, including food hazards that occur before, during, and after harvest; (3) List the critical limits that shall be met at each of the critical control points; (4) List the procedures, and the frequency with which they are to be performed, that will be used to monitor each of the critical control points to ensure compliance with the critical limits; (5) Include any corrective action plans that have been developed in accordance with Section 120.10(a), and that are to be followed in response to deviations from critical limits at critical control points; (6) List the validation and verification procedures, and the frequency with which they are to be performed, that the processor will use in accordance with Sec. 120.11; and (7) provide for a recordkeeping system that documents the monitoring of the critical control points in accordance with Section 120.12. The records shall contain the actual values and observations obtained during monitoring. (c) Sanitation. Sanitation controls may be included in the HACCP plan. However, to the extent that they are monitored in accordance with Section 120.6, they are not required to be included in the HACCP plan.

SECTION 120.9 LEGAL BASIS

Failure of a processor to have and to implement a Hazard Analysis and Critical Control Point (HACCP) system that complies with Sections 120.6, 120.7, and 120.8, or otherwise to operate in accordance with the requirements of this part, shall render the juice products of that processor adulterated under section 402(a)(4) of the Federal Food, Drug, and Cosmetic Act. Whether a processor's actions are consistent with ensuring the safety of juice will be determined through an evaluation of the processor's overall implementation of its HACCP system.

SECTION 120.10 CORRECTIVE ACTIONS

Whenever a deviation from a critical limit occurs, a processor shall take corrective action by following the procedures set forth in paragraph (a) or paragraph (b) of this section. (a) Processors may develop written corrective action plans, which become part of their HACCP plans in accordance with Sec. 120.8(b)(5), by which processors predetermine the corrective actions that they will take whenever there is a deviation from a critical limit. A corrective action plan that is appropriate for a particular deviation is one that describes the steps to be taken and assigns responsibility for taking those steps, to ensure that: (1) No product enters commerce that is either injurious to health or is otherwise adulterated as a result of the deviation; and (2) The cause of the deviation is corrected. (b) When a deviation from a critical limit occurs, and the processor does not have a corrective action plan that is appropriate for that deviation, the processor shall: (1) Segregate and hold the affected product, at least until the requirements of paragraphs (b)(2) and (b)(3) of this section are met; (2) Perform or obtain a review to determine the acceptability of the affected product for distribution. The review shall be performed by an individual or individuals who have adequate training or experience to perform such review; (3) Take corrective action, when necessary, with respect to the affected product to ensure that no product enters commerce that is either injurious to health or is otherwise adulterated as a result of the deviation; (4) Take corrective action, when necessary, to correct the cause of the deviation; and (5) Perform or obtain timely verification in accordance with Section 120.11, by an individual or individuals who have been trained in accordance with Section 120.13, to determine whether modification of the HACCP plan is required to reduce the risk of recurrence of the deviation, and to modify the HACCP plan as necessary. (c) All corrective actions taken in accordance with this section shall be fully documented in records that are subject to verification in accordance with Section 120.11(a)(1)(iv)(B) and the recordkeeping requirements of Section 120.12.

Section 120.11 Verification and validation

(a) Verification. Each processor shall verify that the Hazard Analysis and Critical Control Point (HACCP) system is being implemented according to design. (1) Verification activities shall include: (i) A review of any consumer complaints that have been received by the processor to determine whether such complaints relate to the performance of the HACCP plan or reveal previously unidentified critical control points; (ii) The calibration of process monitoring instruments; (iii) At the option of the processor, the performance of periodic end-product or in-process testing; except that processors of citrus juice that rely in whole or in part on surface treatment of fruit shall perform end-product testing in accordance with Section 120.25. (iv) A review, including signing and dating, by an individual who has been trained in accordance with Section 120.13, of the records that document: (A) The monitoring of critical control points. The purpose of this review shall be, at a minimum, to ensure that the records are complete and to verify that the records document values that are within the critical limits. This review shall occur within 1 week (7 days) of the day that the records are made; (B) The taking of corrective actions. The purpose of this review shall be, at a minimum, to ensure that the records are complete and to verify that appropriate corrective actions were taken in accordance with Section 120.10. This review shall occur within 1 week (7 days) of the day that the records are made; and (c) The calibrating of any process monitoring instruments used at critical control points and the performance of any periodic end product or in process testing that is part of the processor's verification activities. The purpose of these reviews shall be, at a minimum, to ensure that the records are complete and that these activities occurred in accordance with the processor's written procedures. These reviews shall occur within a reasonable time after the records are made; and (v) The following of procedures in Section 120.10 whenever any verification procedure, including the review of consumer complaints, establishes the need to take a corrective action; and (vi) Additional process verification if required by Section 120.25. (2) Records that document the calibration of process monitoring instruments, in accordance with paragraph (a)(1)(iv)(B) of this section, and the performance of any periodic end-product and in-process testing, in accordance with paragraph (a)(1)(iv)(c) of this section, are subject to the recordkeeping requirements of Section 120.12. (b) Validation of the HACCP plan. Each processor shall validate that the HACCP plan is adequate to control food hazards that are reasonably likely to occur; this validation shall occur at least once within 12 months after implementation and at least annually thereafter or whenever any changes in the process occur that could affect the hazard analysis or alter the HACCP plan in any way. Such changes may include changes in the following: raw materials or source of raw materials; product formulation; processing methods or systems, including computers and their software; packaging; finished product distribution systems; or the intended use or consumers of the finished product. The validation shall be performed by an individual or individuals who have been trained in accordance with Section 120.13 and shall be subject to the recordkeeping requirements of Section 120.12. The HACCP plan shall be modified immediately whenever a validation reveals that the plan is no longer adequate to fully meet the requirements of this part. (c) Validation of the hazard analysis. Whenever a juice processor has no HACCP plan because a hazard analysis has revealed no food hazards that are reasonably likely to occur, the processor shall reassess the adequacy of that hazard analysis whenever there are any changes in the process that could reasonably affect whether a food hazard exists. Such changes may include changes in the following: raw materials or source of raw materials; product formulation; processing methods or systems, including computers and their software; packaging; finished product distribution systems; or the intended use or intended consumers of the finished product. The validation of the hazard analysis shall be performed by an individual or individuals who have been trained in accordance with Section 120.13, and, records documenting the validation shall be subject to the recordkeeping requirements of Section 120.12.

SECTION 120.12 RECORDS

(a) Required records. Each processor shall maintain the following records documenting the processor's Hazard Analysis and Critical Control Point (HACCP) system: (1) Records documenting the implementation of the sanitation standard operating procedures (SSOPs) (see Sec. 120.6); (2) The written hazard analysis required by Sec. 120.7; (3) The written HACCP plan required by Sec. 120.8; (4) Records documenting the ongoing application of the HACCP plan that include: (i) Monitoring of critical control points and their critical limits, including the recording of actual times, temperatures, or other measurements, as prescribed in the HACCP plan; and (ii) Corrective actions, including all actions taken in response to a deviation; and (5) Records documenting verification of the HACCP system and validation of the HACCP plan or hazard analysis, as appropriate. (b) General requirements. All records required by this part shall include: (1) The name of the processor or importer and the location of the processor or importer, if the processor or importer has more than one location; (2) The date and time of the activity that the record reflects, except that records required by paragraphs (a)(2), (a)(3), and (a)(5) of this section need not include the time; (3) The signature or initials of the person performing the operation or creating the record; and (4) Where appropriate, the identity of the product and the production code, if any. Processing and other information shall be entered on records at the time that it is observed. The records shall contain the actual values and observations obtained during monitoring. (c) Documentation. (1) The records in paragraphs (a)(2) and (a)(3) of this section shall be signed and dated by the most responsible individual onsite at the processing facility or by a higher-level official of the processor. These signatures shall signify that these records have been accepted by the firm. (2) The records in paragraphs (a)(2) and (a)(3) of this section shall be signed and dated: (i) Upon initial acceptance; (ii) Upon any modification; and (iii) Upon verification and validation in accordance with Section 120.11. (d) Record retention. (1) All records required by this part shall be retained at the processing facility or at the importer's place of business in the United States for, in the case of perishable or refrigerated juices, at least 1 year after the date that such products were prepared, and for, in the case of frozen, preserved, or shelf stable products, 2 years or the shelf life of the product, whichever is greater, after the date that the products were prepared. (2) Offsite storage of processing records required by paragraphs (a)(1) and (a)(4) of this section is permitted after 6 months following the date that the monitoring occurred, if such records can be retrieved and provided onsite within 24 hours of request for official review. Electronic records are considered to be onsite if they are accessible from an onsite location and comply with paragraph (g) of this section. (3) If the processing facility is closed for a prolonged period between seasonal packs, the records may be transferred to some other reasonably accessible location at the end of the seasonal pack but shall be immediately returned to the processing facility for official review upon request. (e) Official review. All records required by this part shall be available for review and copying at reasonable times. (f) Public disclosure. (1) All records required by this part are not available for public disclosure unless they have been previously disclosed to the public, as defined in Section 20.81 of this chapter, or unless they relate to a product or ingredient that has been abandoned and no longer represent a trade secret or confidential commercial or financial information as defined in Section 20.61 of Code of Federal Regulations. (2) Records required to be maintained by this part are subject to disclosure to the extent that they are otherwise publicly available, or that disclosure could not reasonably be expected to cause a competitive hardship, such as generic type HACCP plans that reflect standard industry practices. (g) Records maintained on computers. The maintenance of computerized records, in accordance with part 11 of the Code of Federal Regulations, is acceptable.

SECTION 120.13 TRAINING

(a) Only an individual who has met the requirements of paragraph (b) of this section shall be responsible for the following functions: (1) Developing the hazard analysis, including delineating control measures, as required by Section 120.7; (2) Developing a Hazard Analysis and Critical Control Point (HACCP) plan that is appropriate for a specific processor, in order to meet the requirements of Section 120.8; (3) Verifying and modifying the HACCP plan in accordance with the corrective action procedures specified in Sec. 120.10(b)(5) and the validation activities specified in Sec. 120.11(b) and (c) and Sec. 120.7; (4) Performing the record review required by Sec. 120.11(a)(1)(iv). (b) The individual performing the functions listed in paragraph (a) of this section shall have successfully completed training in the application of HACCP principles to juice processing at least equivalent to that received under standardized curriculum recognized as adequate by the Food and Drug Administration, or shall be otherwise qualified through job experience to perform these functions. Job experience may qualify an individual to perform these functions if such experience has provided knowledge at least equivalent to that provided through the standardized curriculum. The trained individual need not be an employee of the processor.

SECTION 120.14 APPLICATION OF REQUIREMENTS TO IMPORTED PRODUCTS

This section sets forth specific requirements for imported juice. (a) Importer requirements. Every importer of juice shall either: (1) Obtain the juice from a country that has an active memorandum of understanding (MOU) or similar agreement with the Food and Drug Administration, that covers the food and documents the equivalency or compliance of the inspection system of the foreign country with the U.S. system, accurately reflects the relationship between the signing parties, and is functioning and enforceable in its entirety; or (2) Have and implement written procedures for ensuring that the juice that such importer receives for import into the United States was processed in accordance with the requirements of this part. The procedures shall provide, at a minimum: (i) Product specifications that are designed to ensure that the juice is not adulterated under section 402 of the Federal Food, Drug, and Cosmetic Act because it may be injurious to health or because it may have been processed under unsanitary conditions; and (ii) Affirmative steps to ensure that the products being offered for entry were processed under controls that meet the requirements of this part. These steps may include any of the following: (A) Obtaining from the foreign processor the Hazard Analysis and Critical Control Point (HACCP) plan and prerequisite program of the standard operating procedure records required by this part that relate to the specific lot of food being offered for import; (B) Obtaining either a continuing or lot specific certificate from an appropriate foreign government inspection authority or competent third party certifying that the imported food has been processed in accordance with the requirements of this part; (c) Regularly inspecting the foreign processor's facilities to ensure that the imported food is being processed in accordance with the requirements of this part; (D) Maintaining on file a copy, in English, of the foreign processor's hazard analysis and HACCP plan, and a written guarantee from the foreign processor that the imported food is processed in accordance with the requirements of this part; (E) Periodically testing the imported food, and maintaining on file a copy, in English, of a written guarantee from the foreign processor that the imported food is processed in accordance with the requirements of this part; or (F) Other such verification measures as appropriate that provide an equivalent level of assurance of compliance with the requirements of this part. (b) Competent third party. An importer may hire a competent third party to assist with or perform any or all of the verification activities specified in paragraph (a)(2) of this section, including writing the importer's verification procedures on the importer's behalf. (c) Records. The importer shall maintain records, in English, that document the performance and results of the affirmative steps specified in paragraph (a)(2)(ii) of this section. These records shall be subject to the applicable provisions of Section 120.12. (d) Determination of compliance. The importer shall provide evidence that all juice offered for entry into the United States has been processed under conditions that comply with this part. If assurances do not exist that an imported juice has been processed under conditions that are equivalent to those required of domestic processors under this part, the product will appear to be adulterated and will be denied entry.

Pathogen Reduction

SECTION 120.20 GENERAL

This subpart augments subpart A of this part by setting forth specific requirements for process controls.

SECTION 120.24 PROCESS CONTROLS

(a) In order to meet the requirements of subpart A of this part, processors of juice products shall include in their Hazard Analysis and Critical Control Point (HACCP) plans control measures that will consistently produce, at a minimum, a 5-log reduction for a period at least as long as the shelf life of the product when stored under normal and moderate abuse conditions, in the pertinent microorganism. For the purposes of this regulation, the "pertinent microorganism" is the most resistant microorganism of public health significance that is likely to occur in the juice. The following juice processors are exempt from this paragraph: (1) A juice processor that is subject to the requirements of part 113 or part 114 of the Code of Federal Regulations; and (2) A juice processor using a single thermal processing step sufficient to achieve shelf-stability of the juice or a thermal concentration process that includes thermal treatment of all ingredients, provided that the processor includes a copy of the thermal process used to achieve shelf-stability or concentration in its written hazard analysis required by Section 120.7. (b) All juice processors shall meet the requirements of paragraph (a) of this section through treatments that are applied directly to the juice, except that citrus juice processors may use treatments to fruit surfaces, provided that the 5-log reduction process begins after culling and cleaning as defined in Section 120.3(a) and (f) and the reduction is accomplished within a single production facility. (c) All juice processors shall meet the requirements of paragraphs (a) and (b) of this section and perform final product packaging within a single production facility operating under current good manufacturing practices. Processors claiming an exemption under paragraph (a)(1) or (a)(2) of this section shall also process and perform final product packaging of all juice subject to the claimed exemption within a single production facility operating under current good manufacturing practices.

SECTION 120.25 PROCESS VERIFICATION FOR CERTAIN PROCESSORS

Each juice processor that relies on treatments that do not come into direct contact with all parts of the juice to achieve the requirements of Section 120.24 shall analyze the finished product for biotype I *Escherichia coli* as follows: (a) One 20 milliliter (ml) sample (consisting of two 10 ml subsamples) for each 1000 gallons of juice produced shall be sampled each production day. If less than 1000 gallons of juice is produced per day, the sample must be taken for each 1000 gallons produced but not less than once every 5 working days that the facility is producing that juice. Each subsample shall be taken by randomly selecting a package of juice ready for distribution to consumers. (b) If the facility is producing more than one type of juice covered by this section, processors shall take subsamples according to paragraph (a) of this section for each of the covered juice products produced. (c) Processors shall analyze each subsample for the presence of *E. coli* by the method entitled "Analysis for *Escherichia coli* in Citrus Juices — Modification of AOAC Official Method 992.30" or another method that is at least equivalent to this method in terms of accuracy, precision, and sensitivity in detecting *E. coli*. This method is designed to detect the presence or absence of *E. coli* in a 20 ml sample of juice (consisting of two 10 ml subsamples). The method is as follows: (1) Sample size. Total — 20 ml of juice; perform analysis using two 10 ml aliquots. (2) Media. Universal Preenrichment Broth (Difco, Detroit, MI), EC Broth (various manufacturers). (3) Method. ColiComplete (AOAC Official Method 992.30 — modified). (4) Procedure. Perform the following procedure two times: (i) Aseptically inoculate 10 ml of juice into 90 ml of Universal Preenrichment Broth (Difco) and incubate at 35 degrees Celsius for 18 to 24 hours. (ii) Next day, transfer 1 ml of preenriched sample into 10 ml of EC Broth, without durham gas vials. After inoculation, aseptically add a ColiComplete SSD disc into each tube. (iii) Incubate at 44.5 degrees Celsius for 18 to 24 hours. (iv) Examine the tubes under longwave ultraviolet light (366 nm). Fluorescent tubes indicate presence of *E. coli*. (v) MUG positive and negative controls should be used as reference in interpreting fluorescence reactions. Use an *E. coli* for positive control and two negative controls — a MUG negative strain and an uninoculated tube media. (d) If either 10 ml subsample is positive for *E. coli*, the 20 ml sample is recorded as positive and the processor shall: (1) Review monitoring records for the control measures to attain the 5-log reduction standard and correct those conditions and practices that are not met. In addition, the processor may choose to test the sample for the presence of pathogens of concern. (2) If the review of monitoring records or the additional testing indicates that the 5-log reduction standard was not achieved (e.g., a sample is found to be positive for the presence of a pathogen or a deviation in the process or its delivery is identified), the processor shall take corrective action as set forth in Section 120.10. (e) If two samples in a series of seven tests are positive for *E. coli,* the control measures to attain the 5-log reduction standard shall be deemed to be inadequate and the processor shall immediately: (1) Until corrective actions are completed, use an alternative process or processes that achieve the 5-log reduction after the juice has been expressed; (2) Perform a review of the monitoring records for control measures to attain the 5-log reduction standard. The review shall be sufficiently extensive to determine that there are no trends towards loss of control; (i) If the conditions and practices are not being met, correct those that do not conform to the HACCP plan; or (ii) If the conditions and practices are being met, the processor shall validate the HACCP plan in relation to the 5-log reduction standard; and (3) Take corrective action as set forth in Sec. 120.10. Corrective actions shall include ensuring no product enters commerce that is injurious to health as set forth in Section 120.10(a)(1).

REFERENCES

Anonymous, Outbreak of *Salmonella* serotype *muenchen* infections associated with unpasteurized orange juice — United States and Canada, June 1999, *MMWR,* 48, 581–585, 1999.
Barker, W.H. and Runte, V., Tomato juice–associated gastroenteritis, Washington and Oregon, *Am. J. Epidemiol.,* 96, 219–226, 1969.
Besser, R.E., Lett, S.M., Weber, J.T., Doyle, M.P., Barrett, T.J., Wells, J.G., and Griffin, P.M., An outbreak of diarrhea and hemolytic uremic syndrome from *Escherichia coli* O157:H7 in fresh-pressed apple cider, *JAMA,* 269, 2217–2220, 1993.
Memorandum of telephone conversation between Mike Cambridge, New York State Health Department, and Debra Street, FDA, January 22, 1997.
Centers for Disease Control (CDC), *Salmonella typhimurium* outbreak traced to a commercial apple cider — New Jersey, MMWR, 24, 87–88, 1975.
Centers for Disease Control (CDC), Poisoning from elderberry juice — California, *MMWR,* 33, 173–174, 1984.
Centers for Disease Control and Prevention (CDC), Cholera associated with imported frozen coconut milk — Maryland, *MMWR,* 40, 844–843, 1991.
Centers for Disease Control and Prevention (CDC), Outbreak of *Escherichia coli* O157:H7 infections associated with drinking unpasteurized commercial apple juice — British Columbia, California, Colorado, and Washington, October 1996, *MMWR,* 45, 875, 1996a.
Centers for Disease Control and Prevention (CDC), Outbreaks of *Escherichia coli* O157:H7 infection and cryptosporidiosis associated with drinking unpasteurized apple cider — Connecticut and New York, *MMWR,* 46, 4–8, 1996b.
Centers for Disease Control and Prevention (CDC), An Outbreak of Norwalk-like Virus Associated with a Juice Processor in Georgia: Possible Environmental Health Antecedents, The Environmental Health Services Branch, National Center for Environmental Health, Centers for Disease Control and Prevention, July 5, 2000.

Codex Alimentarius Commission, Recommended International Code of Practice: General Requirements (Food Hygiene) [CAC/RCP 1–1969, Rev. 3 (1997), 2nd ed.], HACCP Annex, Food and Agriculture Organization of the United Nations, World Health Organization, Rome.

Cook, K.A., EPI-AID 95–62 Trip Report: Outbreak of *Salmonella hartford* Infections Among Travelers to Orlando, Florida, Centers for Disease Control and Prevention Memorandum, October 1, 1995.

Griffin, P.M., Report of O157:H7 Outbreaks Caused by Juices, Current Science and Technology on Fresh Juices, Transcript of Public Meeting, vol. 1, 15–27, Dec. 16 and 17, 1996.

Memorandum of telephone conversation between Susan Karam, Ohio State Health Department, and Debra Street, FDA, Jan. 21, 1997.

FDA, Health hazard evaluation, classification, and FDA Enforcement Report for recall #F-338/339–8, July 6, 7, and 27, 1988a.

FDA, Health hazard evaluation, classification, and FDA Enforcement Report for firm-initiated recall #F-346–8, July 14, 20, and 27, 1988b.

FDA, Health hazard evaluation, classification, and FDA Enforcement Report for firm-initiated recall #F-68–9, November 16, 1988, and December 6 and 14, 1988c.

FDA, Health hazard evaluation, classification, and FDA Enforcement Report for recall #F-092/093–0, October 19, 1989, and November 1, 1989.

FDA, Health hazard evaluation, classification, and FDA Enforcement Report for recall #F-523–0, June 27, 1990, and July 10 and 18, 1990.

FDA, Health hazard evaluation, classification, and FDA Enforcement Report for firm-initiated recall #F-400/421–1, May 23, 1991, and June 6 and 19, 1991a.

FDA, Health hazard evaluation, classification, and FDA Enforcement Report for firm-initiated recall #F-427–1, June 3, 11, and 19, 1991b.

FDA, Health hazard evaluation, classification, and FDA Enforcement Report for recall #F-529–1, September 4, 5, and 18, 1991c.

FDA, Health hazard evaluation, classification, and FDA Enforcement Report for recall #F-364–2, June 23, 1992, and July 2, 1992a.

FDA, Health hazard evaluation, classification, and FDA Enforcement Report for recall #F-421–2, July 8, 10, and 22, 1992b.

FDA, Health hazard evaluation, classification, and FDA Enforcement Report for firm-initiated recall #F-411/420–2, July 14 and 30, 1992, and August 12, 1992c.

FDA, Health hazard evaluation, classification, and FDA Enforcement Report for recall #F-285–3, April 1 and 21, 1993a.

FDA, Health hazard evaluation, classification, and FDA Enforcement Report for firm initiated recall #F-319–3, May 4 and 19, 1993b.

FDA, Health hazard evaluation, classification, and FDA Enforcement Report for recall #F-250–4, January 13 and 27, 1994, and February 9, 1994a.

FDA, Health hazard evaluation, classification, and FDA Enforcement Report for firm-initiated recall #F-781–4, August 9, 10, and 24, 1994b.

FDA, Health hazard evaluation, classification, and FDA Enforcement Report for firm-initiated recall #F-189/190–5, January 6 and 25, 1995a.

FDA, Health hazard evaluation, classification and FDA Enforcement Report for recall #F-665–5, May 16, 18, and 31, 1995b.

FDA, Pesticide program residue monitoring 1994, *J. AOAC International,* vol. 78, September/October 1995c.

FDA, Health hazard evaluation, classification, and FDA Enforcement Report for firm-initiated recall #F-584/596–6, June 6, 7, and 26, 1996a.

FDA, Health hazard evaluation, classification, and FDA Enforcement Report for recall #F-072-7, November 14 and 20, 1996b.
FDA, Health hazard evaluation, classification, and FDA Enforcement Report for recall #F-073-7, November 14 and 20, 1996c.
FDA, Pesticide Program Residue Monitoring 1995, 1996d.
FDA, Health hazard evaluations, classification, and FDA Enforcement Report for recall #F-492-7, June 27, 1997, and July 2 and 16, 1997a.
FDA, Health hazard evaluation, classification, and FDA Enforcement Report for firm-initiated recall #F-680-7, August 12 and 27, 1997b.
FDA, Health hazard evaluation, classification, and Enforcement Report for recall #F-406-7, May 1 and 7, 1997c.
FDA Enforcement Reports and HHS press release for recall #F-660/661-9, #F-662-9, #F-089-0., September 1 and 15, 1999, November 16 and 1, 1999, and February 2, 2000.
FDA, DHHS, Hazard Analysis and Critical Control Point (HACCP); Procedures for the Safe and Sanitary Processing and Importing of Juice, Proposed Rule, 21 CFR part 120, 63 FR 20450–20486, Apr. 24, 1998.
FDA, DHHS, Hazard Analysis and Critical Control Point (HACCP); Procedures for the Safe and Sanitary Processing and Importing of Juice, Final Rule, 21 CFR part 120, 66 FR 6138–6202, Jan. 19, 2001.
FDA, HHS, FDA Announces Voluntary Nationwide Recall of Certain Frozen Mamey Brands, *HHS News,* March 8, 1999.
Memorandum of telephone conversation between Roberta Hammond, Florida State Health Department, and Debra Street, FDA, Jan. 21, 1997.
Memorandum of telephone conversation between Dr. K. Hendricks, Texas State Health Department, and Debra Street, FDA, Jan. 16, 1997.
Millard, P.S., Gensheimer, K.R., Addiss, D.G., Sosin, D.M., Beckett, G.A., Houck-Jankoski, A., and Hudson, A., An outbreak of cryptosporidiosis from fresh-pressed apple cider, *JAMA,* 272, 1592–1596, 1994.
NACMCF, Hazard Analysis and Critical Control Point principles and application guidelines, adopted August 14, 1997, *J. Food Prot.,* 61, 762–775, 1998.
NACMCF, NACMCF Recommendations on Fresh Juice, April 9, 1997.
National Academy of Sciences (NAS), National Research Council (NRC), *An Evaluation of the Role of Microbiological Criteria for Foods and Food Ingredients,* pp. 41–54, 308–335, National Academy Press, Washington, D.C., 1985.
OSDH, *E. coli* Health Warning Issued in Northeastern Oklahoma. Oklahoma State Department of Health News Release, Oct. 13, 1999.
FDA memorandum from Robert L. Racer, April 21, 2000, concerning April 20 recall of all fresh, unpasteurized citrus juice products because of possible health risks.
Schmelzer, L.L., Gates, J.M., Redfearn, M.S., and Tabershaw, I.R., Gastroenteritis from an orange juice preparation II, Field and laboratory investigation, *Arch. Environ. Health,* 15, 78–82, 1967.
Memorandum of telephone conversation between Pam Shillam, Colorado State Health Department, and Debra Street, FDA, Jan. 17, 1997.
State of Florida, Food Analysis Report for Laboratory No. 98/FO-07398, Oct. 27, 1998; Associated Press, State issues warning about tainted juice, 1998; Garcia, L.M., Traces of bacteria found in juice from local farm, *The Herald* (Miami), Oct. 25, 1998.
Steele, B.T., Murphy, N., Arbus, G.S., and Rance, C.P., An outbreak of hemolytic uremic syndrome associated with ingestions of fresh apple juice, *J. Pediatr.,* 101, 963–965, 1982.

Tabershaw, I.R., Schmelzer, L.L., and Bruyn, H.B., Gastroenteritis from an orange juice preparation I, Clinical and epidemiological aspects, *Arch. Environ. Health,* 15, 72–77, 1967.

Trucksess, M.W., memorandum to Martin J. Stutsman on patulin — 97–688–239, apple juice concentrate, Mar. 31, 1997.

Wagstaff, J., memoranda to M.J. Stutsman and T.C. Troxell on the hazard of patulin in apple juice concentrate sample 97–688–239, April 2, 1997, and July 18, 1997.

Memorandum of telephone conversation between Patty Walker, Washington State Health Department, and Debra Street, FDA, Jan. 15, 1997.

Whatcom County (Washington) Health Department, Summary of a Suspected Outbreak of *E. coli* O157:H7 Associated with Consumption of Unpasteurized Apple Cider, 2 pages, 1996.

Williams, R. et al., unpublished data, 1997.

7 HACCP: An Applied Approach

Todd Konietzko

CONTENTS

HACCP to Control Hazards 158
What Are the Benefits of a HACCP System? 158
Hazard Components 159
Supporting Programs (Prerequisites/Good Manufacturing Practices)......159
Definitions..161
Development of a HACCP Plan..162
    Assemble the HACCP Team 162
    Describe the Food and Its Distribution 162
    Describe the Intended Use and Consumers of the Food 162
    Develop a Flow Diagram that Describes the Process...................163
    Verify the Flow Diagram..163
The Principles of HACCP 163
    Principle 1 — Conduct a Hazard Analysis..................................164
        Examples of Questions to Be Considered When
        Conducting a Hazard Analysis 164
    Principle 2 — Determining Critical Control Points (CCPs) 167
    Principle 3 — Establishing Critical Limits...................................168
    Principle 4 — Establishing Monitoring Procedures 168
    Principle 5 — Establish Corrective Actions 169
        Exceeding the Limit..169
    Principle 6 — Establishing Verification Procedures.....................170
        Validating the HACCP Plan 171
    Principle 7 — Records 171

Juice product safety has become more important with recent foodborne outbreaks that have been linked to juice products. A lack of programs related to food safety may result in personal injury to consumers and economic loss to the manufacturer. However, manufacturing is often the last opportunity to ensure safety prior to use by the consumer. The responsibility for ensuring the safety of juice products rests with the end manufacturer of the product.

Biological hazards are the primary food safety concern; physical and chemical hazards are also a risk. Injury to the consumer can exist in many forms such as physical injury (e.g., cut lip or tongue), allergic reactions (e.g., to a chemical contaminant), illness (food poisoning), and possibly even death.

The U.S. Food and Drug Administration's (FDA) final rule on the use of the Hazard Analysis and Critical Control Point (HACCP) system for ensuring the safety of juice comes after a number of foodborne illness outbreaks that have been associated with juice products during the last several years. Outbreaks have been associated with apple juice and the organism *E. coli* O157:H7, and citrus juice and *Salmonella* spp. An outbreak associated with apple juice resulted in over 70 illnesses and one death, and outbreak(s) linked to citrus products have resulted in over 500 reported illnesses, as well as one reported death. Foodborne infections are dangerous for the elderly, the young, and those whose immune systems have been weakened. The FDA has estimated that over 16,000 cases of juice-related illness occur every year. The FDA believes that action taken due to the final juice HACCP rule will prevent an estimated 6000 of those illnesses per year.

HACCP TO CONTROL HAZARDS

Juice products, because of their added sugar, nutrients, and an increase in the pH of the products, may provide an ideal medium for the growth of bacteria, including pathogenic bacteria.

A key juice HACCP requirement is the 5-log reduction requirement for the microbe of concern, which has been identified as the pathogen that has had a history of outbreaks in a specific juice. This requirement can be satisfied by one or a series of measures or steps that can reduce pathogenic bacterial counts by at least 5 logs. The FDA has recognized pasteurization as a method for meeting the 5-log reduction requirement. Also, part of the rule states that the reduction must occur in one processing facility. Measures taken prior to the receipt of the juice cannot be included in the total 5-log reduction. In addition, the final processor must have assurance that its HACCP plan results in the elimination of the pathogen of concern.

WHAT ARE THE BENEFITS OF A HACCP SYSTEM?

The primary concern of a HACCP system and its supporting programs is food safety. The manufacturer who can successfully implement the food safety program may also see other benefits, including the following:

1. Decrease in consumer complaints, including physical, illness, and quality complaints
2. Reduction of holds of product in the manufacturing facility
3. Consumer confidence in the brand, leading to retention or increase in market share

A successful HACCP plan must consider any potential (physical, biological, chemical) hazard that is reasonably likely to occur. Potential hazards associated with incoming fruit, foreign material, and added ingredients to make the finished product must all be included in a complete hazard analysis. During the steps of the process, a control measure should be identified for each potential hazard.

HAZARD COMPONENTS

Hazards that need to be considered in each facility's hazard analysis are those that are reasonably likely to occur in the facility. Consideration should be given to ingredients, each process step within the facility, and finished product packaging and final storage. The facility's hazard analysis will be unique to the particular processing plant. One thing to remember when performing the hazard analysis is that hazards are defined within HACCP as they relate to product safety. The hazards included in the analysis should be only those that are reasonably likely to occur in the process being evaluated. The FDA final regulation on juice HACCP has identified nine hazards that must be addressed in the written plant hazard analysis. They are:

1. Microbial contamination
2. Parasites
3. Chemical contamination
4. Decomposition
5. Unlawful pesticide residues
6. Natural toxins
7. Unapproved food or color additives
8. Physical hazards
9. Undeclared ingredients, such as allergens

SUPPORTING PROGRAMS (PREREQUISITES/GOOD MANUFACTURING PRACTICES)

The manufacturing plant's HACCP program cannot and is not meant to be a stand-alone program. Instead, it is part of a larger system including company Good Manufacturing Practices (GMPs), plant Standard Operating Procedures (SOPs), and Sanitary Standard Operating Procedures (SSOPs). Prerequisite programs are procedures used to control the plant operating conditions and employee practices, and ultimately contribute to the production of a safe, quality food product.

The effectiveness and existence of prerequisite programs should be assessed during the design and implementation of the HACCP plan. Prerequisite programs, like the HACCP program itself, should be documented and audited regularly. Prerequisite programs should be managed separately from the HACCP program. However, it may be necessary to include parts of a prerequisite program in the HACCP plan. One possible example would be a preventative maintenance procedure for equipment to avoid unexpected down time. During the development of a HACCP plan, it may be decided that routine maintenance and calibration of a thermometer on a pasteurization system should be included in the plan as an activity of verification. This then would ensure that the product passing through the system met the designated temperature necessary for the production of a safe food.

In the final juice HACCP regulation, FDA has required that the operating plant have in place eight SSOPs. These eight SSOPs can be classified as prerequisite programs. The eight mandatory SSOPs are:

1. Safety of water that comes into contact with food or food contact surfaces or that is used in the manufacture of ice
2. Condition and cleanliness of food contact surfaces, including utensils, gloves, and outer garments
3. Prevention of cross contamination from unsanitary objects to food, food packaging materials, and other food contact surfaces, including utensils, gloves, and outer garments, and from raw product to processed product
4. Maintenance of handwashing, hand sanitizing, and toilet facilities
5. Protection of food, food contact packaging material, and food contact surfaces from adulteration with lubricants, fuel, pesticides, cleaning compounds, sanitizing agents, condensate, and other physical, chemical, or biological contaminates
6. Proper labeling, storage, and use of toxic compounds
7. Control of employee health conditions that could result in the microbiological contamination of food, food packaging material, and food contact surfaces
8. Exclusion of pests from the food plant

The FDA final regulation does not state that these eight SSOPs have to be written programs; it just requires documentation that the eight SSOPs are being controlled.

HACCP: An Applied Approach

End product sampling and testing by the processor are not required. If testing is included as part of the verification step in the HACCP plan, then referenced industry-testing records will have to be made available to a regulatory agency during an audit/inspection of the facility.

DEFINITIONS

In order to develop a HACCP plan, certain terms must be defined and understood:

CCP Decision Tree   A series of questions to determine whether a control point is a critical control point (CCP).
Control   To manage the conditions of an operation to maintain compliance with established criteria.
Control Measure   An action or activity that can be used to prevent, eliminate, or reduce a food safety hazard to an acceptable level.
Corrective Action   A written set of procedures that are followed when a deviation occurs.
Critical Control Point (CCP)   A step at which control can be applied and is essential to prevent, eliminate, or reduce a food safety hazard to acceptable levels.
Critical Limit   A maximum and/or limited value to which a biological, chemical, or physical parameter must be controlled at a CCP to prevent, eliminate, or reduce to an acceptable level the occurrence of a food safety hazard.
Deviation   Failure to meet a critical limit.
HACCP plan   The written document based on the principles of HACCP that explains procedures to be followed to ensure control is based on the seven principles of HACCP.
HACCP team   The group of people in the processing facility who are responsible for the development, implementation, and maintenance of the HACCP system.
Hazard   A biological, chemical, or physical agent that is reasonably likely to cause illness or injury in the absence of control.
Hazard analysis   The process of collecting and reviewing information and data on hazards associated with the food under consideration, to decide which are significant and must be addressed in the HACCP plan.
Prerequisite Programs   Procedures and policies, including GMPs, SOPs, and SSOPs, that address operational conditions, which provide for the foundation of the HACCP system.

Validation   The element of verification, which is focused on collecting and evaluating scientific and technical information to determine if the HACCP plan, when properly implemented, will effectively control the identified hazards.

Verification   Activities other then normal monitoring that determine the validity of the HACCP plan and that the system is operating according to the plan.

DEVELOPMENT OF A HACCP PLAN

Development of a facility's HACCP plan is not a task that can be accomplished by one individual. First and foremost, it is necessary to get the needed commitment from all management levels of the facility. Five preliminary tasks must be accomplished before the plan can be developed. They are:

Assemble the HACCP team.
Describe the food (produced) and its distribution.
Describe the intended use and consumers of the food.
Develop a flow diagram that describes the flow in the facility.
Verify the flow diagram.

ASSEMBLE THE HACCP TEAM

The members of the HACCP team are individuals who have knowledge and expertise appropriate to the facility's products and processes. This team is then responsible for the development of the HACCP plan. The team should be cross-functional and include employees from production, sanitation, quality assurance/control, microbiology, engineering, and management. Occasionally, the HACCP team may need assistance from outside consultants (experts) who are knowledgeable about possible chemical, physical, and biological hazards associated with the process or final product.

DESCRIBE THE FOOD AND ITS DISTRIBUTION

The first task of the HACCP team is to describe the food. This task consists of a general description of the food, ingredients, processing, and final packaging. The distribution method should be described, along with information on whether the food is to be distributed frozen, refrigerated, or at ambient temperatures.

DESCRIBE THE INTENDED USE AND CONSUMERS OF THE FOOD

In this area, the team shall describe the expected use of the food and the intended users, whether general public or a particular segment of the population (e.g., infants, elderly people, etc.).

HACCP: An Applied Approach

Develop a Flow Diagram that Describes the Process

The purpose of the flow diagram is to provide an outline of the steps involved in the process. The flow diagram must cover all the steps in the process, beginning at the point where ingredients enter the plant, and including all steps under the direct control of the establishment. The plan can also include steps in the process beyond the direct control of the plant if the hazards included are significant and can only be controlled within the plant. A block flow diagram is sufficient for documentation of the flow; complex engineering drawings are not necessary. Points to consider when establishing the flow diagram include:

All process steps where raw materials, ingredients, and packaging are used
All process steps in production
Product recycle/rework and waste areas
Ingredient and product storage and distribution

Verify the Flow Diagram

The HACCP team should take the developed flow diagram out to the processing operation to verify the accuracy and completeness of the developed diagram. Changes to the flow diagram should be made so that the diagram accurately represents the actual process steps in the facility. Any changes to the flow diagram should be documented and dated as part of the HACCP plan. Periodically thereafter, the flow diagram should be taken to the production floor to ensure the diagram's accuracy.

THE PRINCIPLES OF HACCP

Once the HACCP team has worked and completed the five preliminary steps of the HACCP system, the team can then begin the process of developing the HACCP plan by utilizing the seven principles of HACCP for establishing an effective HACCP program. The seven steps are as follows:

1. Conduct a hazard analysis.
2. Determine critical control points (CCPs).
3. Establish critical limits.
4. Establish monitoring procedures.
5. Establish corrective actions.
6. Establish verification procedures.
7. Establish recordkeeping and documentation procedures.

During the application of the seven principles of HACCP, the HACCP plan summary table will be useful as the HACCP team's working document.

Principle 1 — Conduct a Hazard Analysis

The purpose of the hazard analysis is to develop a list of hazards. A hazard is any biological, chemical, or physical agent that is reasonably likely to cause illness or injury in the absence of control. As the HACCP team conducts the hazard analysis and identifies the appropriate control measures (for the process/product being evaluated), the team should begin to develop a list of hazards. Hazards not reasonably likely to occur do not require further consideration within the HACCP plan. A thorough hazard analysis is essential to preparing an effective HACCP plan. If the hazard analysis is not done correctly, and the hazards that warrant control within the plan are not identified, the plan will not be effective regardless of how well it is followed.

In the hazard analysis, three objectives are accomplished:

Hazards and associated control measures are identified.
The analysis may identify a need for changes in the process or product so that product safety is further ensured or improved.
The analysis provides a basis for determining CCPs in Principle 2.

There are two stages involved in conducting a hazard analysis. The first is hazard identification. During this stage, the HACCP team reviews ingredients used, activities conducted at each step in the process, and equipment used in producing the final product, along with methods of storage and distribution. Also, the intended use and consumers of the product are evaluated. Based on the review, the team develops a list of potential biological, chemical, or physical hazards, which may be introduced, increased, or controlled at each step in the production process.

Examples of Questions to Be Considered When Conducting a Hazard Analysis

The hazard analysis consists of asking questions that are appropriate to the process under consideration. The purpose of the questions is to assist in identifying potential hazards. For example:

1. Ingredients
    a. Does the food contain any sensitive ingredients that may present microbiological, chemical, or physical hazards?

HACCP: An Applied Approach

     b. Are potable water, ice and/or steam used in the formulation or handling of the food?
     c. What are the sources?
2. Intrinsic factors — physical characteristics and composition (i.e., pH, water activity, preservatives) of the food during and after processing
     a. What hazards may result if the food composition is not controlled?
     b. Does the food permit the survival or multiplication of pathogens and/or toxin formation in the food during processing?
     c. Will the food permit the survival or multiplication of pathogens and/or toxin formation during subsequent steps in the food chain?
     d. Are there other similar products on the market? What has been the safety record of those products? What hazards have been associated with the products?
3. Procedures used for processing
     a. Does the process include a controllable processing step that destroys pathogens? Which pathogens have both vegetative cells and spores, and have they been included?
     b. If the product is subject to recontamination between processing (e.g., pasteurization) and packaging, which chemical, physical, or biological hazards are likely to occur?
4. Microbial content of the food
     a. What is the normal microbial content of the food?
     b. Does the microbial population change during the time the food is stored prior to consumption?
     c. Does the change in the microbial population alter the safety of the food?
     d. Do the answers to these questions indicate a high likelihood of certain biological hazards?
5. Facility design
     a. Does the layout of the facility provide separation of raw materials from ready-to-eat (RTE) foods if this is important in food safety?
     b. Is positive air pressure maintained in product packaging areas? Is this essential?
     c. Is the traffic pattern in the facility for people and equipment a source of contamination?
6. Equipment design and use
     a. Will equipment provide the time–temperature control that is necessary for safe food?
     b. Is equipment sized for the volume being processed?

c. Can equipment be controlled so that the variation in performance will be within the tolerances required for the production of safe food?
d. Is equipment reliable or prone to breakdowns?
e. Is equipment designed for easy cleaning and sanitizing?
f. Is there a chance for product contamination with hazardous substances (i.e., glass, lubricants)?
g. What product safety devices are used to enhance safety?
   Metal detectors
   Magnets
   Sifters
   Screens
   Bone removal systems
   Filters
   Thermometers
h. To what degree does normal equipment wear affect the possible occurrence of a physical hazard (e.g., metal and glass)?
i. Are allergen protocols needed in using equipment for different products?

7. Packaging
   a. Does the method of packaging affect the multiplication of pathogens and/or the formation of toxins?
   b. Is the package clearly labeled for the required storage conditions for safety (e.g., Keep refrigerated, frozen until use)?
   c. Does the package include instructions for the safe handling and preparation of the food by the end consumer?
   d. Is the package resistant to damage, preventing the introduction of microbial contamination?
   e. Are tamper evident packaging features used?
   f. Is each package and case legible and coded correctly?
   g. Does each package have the proper label?
   h. Are allergens in the product properly labeled in the ingredient listing on the label?

8. Sanitation
   a. Can sanitation have an impact upon the safety of the food being processed?
   b. Can the facility and equipment be easily cleaned and sanitized to permit the safe handling of food?
   c. Is it possible to provide sanitary conditions consistently and adequately to ensure safe foods?

9. Employee health, hygiene, and education

HACCP: An Applied Approach

   a. Can employee health and hygiene practices impact upon the safety of the food being processed?
   b. Do the employees understand the process and the factors they must control to ensure the preparation of safe food?
   c. Will employees inform management of a problem that could impact upon the safety of the food?
10. Conditions of storage between packaging and the end user
   a. What is the likelihood that the food will be improperly stored at the wrong temperature?
   b. Would an improper storage error lead to a microbiologically unsafe food?
11. Intended use
   a. Will the consumer heat the food?
   b. Will there likely be any leftovers?
12. Intended consumer
   a. Is the food intended for the general public?
   b. Is the food intended for consumption by a population with increased susceptibility to illness (i.e., infants, elderly, immunocompromised individuals)?

PRINCIPLE 2 — DETERMINING CRITICAL CONTROL POINTS (CCPs)

A critical control point (CCP) is defined as a step at which control is applied to prevent, eliminate, or reduce to an acceptable level a chemical, physical, or biological hazard. The hazard analysis when conducted properly under Principle 1 should identify hazards that need to be controlled. Prerequisite CGMPs and mandatory SSOPs may also be used to control many of the identified hazards. A hazard that is not controlled through these programs must be controlled at CCPs. CCPs can vary depending on the hazard analysis, plant, product, and production methods.

The HACCP team, through the information developed during the hazard analysis, should be able to identify which steps in the process are CCPs. Each CCP can be further identified by the use of a CCP decision tree. The CCP decision tree can be useful for help in determining if a particular step is a CCP for an identified hazard; however, it is merely a tool and not a mandatory element of HACCP.

Facilities that process the same type of food products can differ in the identified risks and hazards as well as the points, steps, or procedures that are designated CCPs. These differences can be due to facility layout, equipment, processing steps, and selection of ingredients.

A HACCP summary table is useful for documentation of a critical control point (see column 1). The hazard that has been identified at the CCP should then be transferred to column 2 of the summary table.

PRINCIPLE 3 — ESTABLISHING CRITICAL LIMITS

Once CCPs have been identified, parameters need to be established to signify whether the control measure at the CCP is "in" or "out" of control. These parameters are defined as the critical limits. Thus, a critical limit is used to distinguish between safe and unsafe operating conditions at the CCP. Operational limits differ from critical limits, in that operational limits are established for reasons other than food safety. Critical limits must be met to ensure the safety of the product being produced. Critical limits can be derived from sources such as regulatory standards, literature searches, studies, and experts. Critical limits are parameters that have been established for control measures and might include:

Temperature	Time
Water activity	pH
Moisture levels	Viscosity
Flow rate	Salt concentration

When a critical limit is not met, it should be indicated to the processing facility that a CCP is out of control and that a potential exists for the development of a health hazard.

PRINCIPLE 4 — ESTABLISHING MONITORING PROCEDURES

Monitoring consists of planned observations or measurements to assess whether a CCP is under control and to produce an accurate record for future use in verification. Correct monitoring has three purposes:

1. Monitoring is essential to the identified product's food safety management in that it tracks the systems operation.
2. Monitoring is used to determine when there is loss of control and a deviation has occurred at a CCP, and when corrective action must be taken.
3. Monitoring provides a written document for use in verification of the facility's HACCP plan.

Because of the consequences of a deviation, monitoring procedures must be effective. Monitoring equipment must be calibrated for accuracy. When it is not possible to monitor a critical limit continuously, it is necessary to establish an effective monitoring schedule that is reliable enough to record that the identified hazard is under control.

Assignment of the responsibility for monitoring is an important consideration for each CCP. Specific assignments depend on the number of CCPs and the critical limit. Individuals who monitor CCPs must:

HACCP: An Applied Approach

Be trained in the technique used to monitor each critical limit
Understand the purpose and importance of monitoring
Have ready access to monitoring activity
Report monitoring activity accurately
Sign or initial the monitoring records

Individual(s) responsible for conducting the monitoring activities should be trained in the specific monitoring activities that are used. Each individual should be trained so that he or she understands that the monitoring and reporting needs to be done accurately and consistently from day to day.

It is also important for the person who is responsible for monitoring to inform management when a process or product fails to meet the critical limits so that proper and immediate corrective actions can be taken.

Once effective monitoring procedures are established, this information can be transferred to columns 4 to 7, and records being monitored can be placed in column 10 on the HACCP plant summary table.

PRINCIPLE 5 — ESTABLISHING CORRECTIVE ACTIONS

Corrective actions are procedures that the operator of the processing facility follows when a deviation occurs. Because of the variations in CCPs and the number of possible deviations, specific corrective action plans are developed for each CCP. Production personnel should have the authority and responsibility to take corrective actions such as shutting down lines and contacting management. Actions taken must demonstrate that the CCP has been brought under control. As part of the HACCP plan, corrective action procedures must be documented.

Written corrective action plans may include the following actions:

Eliminate the actual or potential hazard created by the deviation.
Develop specific corrective actions for each CCP.
    Halt production of the product.
    Isolate the affected product.
    Return processes to control.
    Determine the disposition of the product.
Ensure safe disposition of the product involved.
Demonstrate that the CCP has been brought under control.
Determine the source to prevent recurrence.

Exceeding the Limit

Decisions must be based on the fact that exceeding the limit may indicate:

Evidence or existence of a direct health hazard
Evidence that a direct heath hazard could develop
Evidence of a CCP not under control

Once corrective actions have been developed, they can be transferred to column 8 on the HACCP summary plan.

Principle 6 — Establishing Verification Procedures

Documentation is critical in the HACCP plan. Verification can be addressed in two ways, first by establishing verification on the HACCP plan summary table and second by establishing internal monitoring verification programs.

Verification is the activity designed to ensure that the monitoring is operating according to the requirements of the program. The format in which verification is determined is up to the HACCP team in terms of what actions will be taken in reference to the CCP(s) and is placed in column 9 of the HACCP plan summary table.

Once verification has been determined, internal verification monitoring programs can be developed.

Examples of internal verification monitoring include:

Appropriate verification monitoring schedules
CCP review at specific frequencies
After the HACCP plan has been modified
When the food has been implicated as a source of foodborne disease
Visual verification of operations to observe if the identified CCP is under control

Verification records should include:

Record status associated with CCP monitoring
Records of direct monitoring of the CCP while in operation
Review of records showing calibration of monitoring equipment to ensure that all devices are calibrated and in working order
Deviations and corrective actions
Training of individuals responsible for monitoring of CCPs
Documentation showing that records have been verified

At this point, if internal verification and monitoring are found unacceptable in resolving a deviation issue, then validation of the plan by the team may be necessary.

HACCP: An Applied Approach

Validating the HACCP Plan

The plan and the entire program should be validated annually and should be conducted by the team. The process to validate the HACCP plan, CCPs, and critical limits involves determining whether the set of controls is adequate to prevent, eliminate, or reduce to an acceptable level identified hazards.

When validating the HACCP plan on an annual basis, the HACCP team should consider reviewing some of the following items:

Effectiveness of the process
Flow diagram accuracy
The hazard analysis
Identification and whether appropriate CCPs have been identified
Justification of critical limits
Effectiveness of monitoring programs and recordkeeping

In addition to the annual review, some situations that require validation include:

New identified hazard associated with the food
Recalls of like/similar foods
Formulation, packaging changes
Equipment changes
Changes in product flow within the plant
Responses to regulatory inspection/rules

Principle 7 — Records

All records used in the HACCP system should be available for review and verification purposes, including:

The HACCP plan
Listing of team and areas of expertise
Description and intended use of food being produced
Flow diagram for the facility and identification of CCPs
Monitoring systems
Corrective action plans for when deviations from the critical limits occur
Verification procedures for the HACCP system
Hazard analysis
Validation procedures
Records for all CCPs

The following records must be available according to 21 CFR Part 120.12, Hazard Analysis and Critical Control Point (HACCP); Procedures for the Safe and Sanitary Processing and Importing of Juice, Final Rule for the Facilities HACCP Plan upon inspection:

Records documenting the implementation of the sanitation standard operating procedures as outlined in Sec. 120.6 of the final regulation, including sanitation controls that address and implement a sanitation standard operating procedure that addresses sanitation conditions before, during, and after processing. The SSOPs shall address:

1. Safety of water that comes into contact with food or food contact surfaces or that is used in the manufacture of ice
2. Condition and cleanliness of food contact surfaces, including utensils, gloves, and outer garments
3. Prevention of cross contamination from unsanitary objects to food, food packaging materials, and other food contact surfaces, including utensils, gloves, and outer garments, and from raw product to processed product
4. Maintenance of handwashing, hand sanitizing, and toilet facilities
5. Protection of food, food contact packaging material, and food contact surfaces from adulteration with lubricants, fuel, pesticides, cleaning compounds, sanitizing agents, condensate, and other physical, chemical, or biological contaminates
6. Proper labeling, storage, and use of toxic compounds
7. Control of employee health conditions that could result in the microbiological contamination of food, food packaging material, and food contact surfaces
8. Exclusion of pests from the food plant

Each processor shall maintain SSOP records that, at a minimum, document the monitoring and corrections as mandated by the final regulation and listed above. These records shall be maintained as outlined in 120.12 (d) in that: All records required by this part shall be retained at the processing facility or at the importer's place of business in the United States for, in the case of perishable or refrigerated juices, at least 1 year after the date that such products were prepared, and for, in the case of frozen, preserved, or shelf stable products, 2 years or the shelf life of the product, whichever is greater, after the date that the products were prepared.

In light of the recent rise in outbreaks of foodborne illness and the occurrence of several outbreaks linked to juice and juice-based products, the beverage industry needs to embrace the concept of HACCP, its programs and documentation, showing that the industry takes these concerns and issues seriously. The industry needs to remember that the first element of a successful HACCP plan is the completion of the seven principles. Along with that, training and documentation must be developed for the principles and kept together in the development and implementation of the program.

8 Essential Elements of Sanitation in the Beverage Industry

Martha Hudak-Roos and Bruce Ferree

CONTENTS

Introduction 175
Why Are There Sanitation Needs? 176
What Are the Current and Future Regulatory Components
of a Sanitation Program? 177
    GMPs 179
    Bottled Water 182
    Juice Regulations 183
    Other Regulations 184
    Summary 184
What Tools Are Used by the Beverage Industry Today in Fulfilling
Its Sanitation Program Needs? 184
    Cleaning and Sanitizing 184
    Procedures 186
    Other Sanitation Elements 187
    Master Sanitation Schedule 188
How Is the Efficacy of the Program Verified? 189
Case Studies for Beverages 190
Conclusion 192

INTRODUCTION

Sanitation is an essential element for any food processor but is especially crucial for beverage processors.

Key elements in sanitation include:

Why are there sanitation needs?
What are the current and future regulatory components of a sanitation program?
What tools are used by the beverage industry today in fulfilling its sanitation program needs?
How is the efficacy of the program verified?

Examples of these key elements will be illustrated with case studies. By following basic principles and incorporating important components such as management commitment, training, and resources into the sanitation program, beverage manufacturers can maintain the efficacy of their sanitation programs, resulting in high-quality, safe beverages that meet reasonable shelf life expectations.

WHY ARE THERE SANITATION NEEDS?

The word *sanitation* is derived from the Latin "sanitas," which means health. Sanitation in the food industry has been applied to the process of creating and maintaining a wholesome environment in which to make safe food. It is a broad-based program — encompassing in today's food industry a large portion of what are considered "prerequisite" or "universal" food safety program elements. These are program elements that are plant wide, as opposed to the Hazard Analysis and Critical Control Point (HACCP) system, which identifies specific process steps as the essential or critical control points. These universal programs are the second level of the food safety pyramid (Figure 8.1), with management commitment as the base and HACCP and continuous quality improvement as the next layer and the pinnacle, respectively.

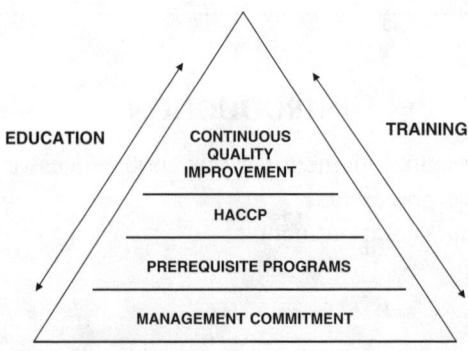

FIGURE 8.1 Food safety and quality pyramid.

Sanitation covers water, air, employee hygiene, equipment cleanliness, equipment maintenance, facility design and condition, pest control, chemical control, and a host of other plant-wide needs. By their very nature, sanitation programs fulfill the moral and legal obligations to create and maintain hygienic practices that keep a facility, equipment, and hence the food clean and wholesome. Companies that have failed in their sanitation programs have gone down in the annals of history and in the stock market. The recalls involving Hudson Industries and Bil Mar Foods could have been somewhat mitigated with enhanced sanitation programs. Outside of the safe food aspects of sanitation, a clean environment can increase product quality and shelf life. Reduced yeast and mold in the product and in the environment are key goals of sanitation, and these organisms affect the quality of products as well.

As a broad-based program within any plant's food safety system, sanitation must have training, education, and documentation as key elements. It is unfortunate that we do not require "certified" sanitation managers in the industry. In many cases, it is difficult to get a sanitation crew together. Sanitation is typically performed during the night shift and involves very inhospitable environments. The members of the sanitation crew are truly the unsung heroes of the food safety system.

WHAT ARE THE CURRENT AND FUTURE REGULATORY COMPONENTS OF A SANITATION PROGRAM?

The basic sanitation regulation for the beverage industry begins with 21 Code of Federal Regulations Part 110, also known as "Current Good Manufacturing Practices." For those beverage manufacturers involved in international trade, hygiene standards are found in the Codex Alimentarius Commission's "General Principles of Food Hygiene." These two documents list comparable considerations in producing safe, wholesome beverages in a sanitary environment.

The Good Manufacturing Practices (GMPs) define the regulatory expectations for sanitation. For a beverage processor, the beginning step in sanitation is the process of identifying how these expectations are met. We suggest a table format such as Table 8.1.

In this table, the left column is the Food and Drug Administration (FDA) provision and the middle column is the GMP text as spelled out in 21 CFR 110. The last column is where the plant/facility can list the procedures or programs that are used to cover the regulatory need. Such a matrix can identify any GMP area that is not fully addressed. It can also help to consolidate procedures, if you find within the matrix many

TABLE 8.1
U.S. FDA GMP Matrix Example

GMP Item	Regulatory Guidance	Food Safety Program
110.10 Personnel	The plant management shall take all reasonable measures and precautions to ensure the following.	
(a) Disease Control	Any person who, by medical examination or supervisory observation, is shown to have, or appears to have, an illness, open lesion, including boils, sores, or infected wounds, or any other abnormal source of microbial contamination by which there is a reasonable possibility of food, food-contact surfaces, or food-packaging materials becoming contaminated, shall be excluded from any operations which may be expected to result in such contamination until the condition is corrected. Personnel shall be instructed to report such health conditions to their supervisors.	
(b) Cleanliness	All persons working in direct contact with food, food-contact surfaces, and food-packaging materials shall conform to hygienic practices while on duty to the extent necessary to protect against contamination of food. The methods for maintaining cleanliness include, but are not limited to: (1) Wearing outer garments suitable to the operation in a manner that protects against the contamination of food, food-contact surfaces, or food-packaging materials. (2) Maintaining adequate personal cleanliness. (3) Washing hands thoroughly (and sanitizing if necessary to protect against contamination with undesirable microorganisms) in an adequate handwashing facility before starting work, after each absence from the work station, and at any other time when the hands may have become soiled or contaminated.	

TABLE 8.1 (CONTINUED)
U.S. FDA GMP Matrix Example

GMP Item	Regulatory Guidance	Food Safety Program
	(4) Removing all insecure jewelry and other objects that might fall into food, equipment, or containers, and removing hand jewelry that cannot be adequately sanitized during periods in which food is manipulated by hand. If such hand jewelry cannot be removed, it may be covered by material which can be maintained in an intact, clean, and sanitary condition and which effectively protects against the contamination by these objects of the food, food-contact surfaces, or food-packaging materials.	
	(5) Maintaining gloves, if they are used in food handling, in an intact, clean, and sanitary condition. The gloves should be of an impermeable material.	

procedures that are addressing the same need. For international companies, the Codex General Principles of Food Hygiene are analogous to the FDA GMPs and are an excellent resource as the base reference for any company's sanitation program.

GMPs

The General Provisions within the GMPs start with one of the most important aspects of hygiene, personnel hygiene. Plant management must have a program in place to ensure that personnel are healthy and not likely to be the source of abnormal microbial contamination. Also, personnel working with food or packaging must maintain a degree of cleanliness that includes clean outer garments, good personal hygiene, and adequate handwashing. Jewelry is not permitted. Gloves, if used, must be intact, clean, and in sanitary condition, as well as made from an impervious material. Hair restraints are required. Storage of personal items is limited to areas other than where food is exposed or equipment and utensils are washed. Personal habits are also regulated: no gum chewing, eating food, drinking beverages, or using tobacco where food is exposed or equipment is washed. Finally, the regulatory guidance concludes with the statement, "…taking necessary precautions to protect against contamination of food, food-contact surfaces, or food packaging materials with microorganisms or foreign substances including, but not limited to, perspiration, hair, cosmetics, tobacco, chemicals, and medicines applied to the skin."

The General Provisions also cover education and training. The guidance suggests that personnel responsible for identifying sanitation failures or food contamination should have a background of education or experience to provide a level of competency necessary for the production of safe food. Food handlers should receive (in the authors' opinion, *must* receive) appropriate training in proper food handling techniques and food protection principles and should be informed of the dangers of poor personal hygiene and unsanitary practices. Supervisors, according to the GMPs, are responsible for ensuring compliance.

The grounds of the manufacturing site should allow for proper storage of equipment and removal of waste. Weeds and grass should be kept cut, and the yard, roads, etc., maintained. Reducing pest harborage and attractants is part of this requirement: drainage and waste disposal must be adequate. All unused process piping must be stored up off the ground. The building also must be under control. Sufficient space for equipment and storage must be available. The design of the facility must be effective to permit proper precautions for reducing the potential for contamination: i.e., location, time, partition, airflow, and enclosed systems. Included in this section of the GMPs is the requirement to protect food in outdoor bulk fermentation vessels by coverage, controlling areas over the tanks, checking on a regular basis for pests, and skimming.

Building Controls go on to cover floors, walls, ceilings (construction, materials, cleanliness), lighting, ventilation, and screening for pests. For sanitary operations, the GMPs require that the building and equipment be maintained in a sanitary condition. Cleaning compounds and sanitizing agents must be safe and adequate for use. (*Note:* the GMPs do not define acceptable cleaning and sanitizing compounds. These are defined in 21 CFR Part 178.) Toxic compounds must be controlled so as not to contaminate food, contact surfaces, or packaging. Pest control is required within this section as well. Sanitary operations also include the cleaning and sanitizing of food contact surfaces.

Sanitary Facilities and Controls covers water supply, plumbing, sewage, and toilets. Processors who use boilers to create steam should become familiar with this section as well as Part 173 Subpart D — Secondary Direct Food Additives, Specific Usage Additives.

Finally, this section covers handwashing. Good handwashing includes hot water, sanitary towels, and refuse receptacles for waste towels. Signs that indicate that hands must be washed and how to wash must be posted and must be understood easily by the staff. Fixtures with foot- or knee-activated pedals that are designed to protect against recontamination of clean hands are also important.

Equipment and Utensils are part of the GMPs. According to the regulation, "The design, construction, and use of equipment and utensils shall preclude the adulteration of food with lubricants, fuel, metal fragments, contaminated water, or any other contaminants. All equipment should be installed and maintained as to facilitate the cleaning of equipment and all adjacent areas." This section covers aspects important to the beverage industry such as smooth welds and the cleanliness of holding, conveying, and manufacturing systems (closed and automated). It covers equipment instrumentation (thermometers, pH, acidity, etc.) and the need to accurately and adequately maintain these devices. This is an important section for review.

Production and Process Controls covers operations. This section has some key elements, including "Appropriate quality-control operations shall be employed to ensure that food is suitable for human consumption and that food packaging materials are safe and suitable" and "Overall sanitation of the plant shall be under the supervision of one or more competent individuals assigned responsibility for this function." There are three areas specific to this section: Raw Materials, Manufacturing Operations, and Warehousing. The Raw Materials section requires that raw materials be inspected, segregated, or otherwise handled to ensure that they are clean and suitable for processing. This section also requires that water used for washing or rinsing be safe and of sanitary quality. Where water is reused for washing, it cannot increase the level of contamination of the food. In today's manufacturing environments, water reuse (recycling) is a key element in food sanitation. The Codex Committee on Food Hygiene has developed a text on the safe reuse of water that can be found on the Codex Web site. Those in the beverage industry must be cognizant of water safety issues, including *Cryptosporidium* (chlorine resistant), *Giardia*, and other protozoa. Rinsing and sanitizing equipment right before use with chlorinated water might not provide for safe food contact surfaces.

Other points of interest within Raw Materials include:

"Material scheduled for rework shall be identified as such."
 Check your rework protocols to ensure that they meet this requirement.
"Liquid materials received and stored in bulk shall be held in a manner that protects against contamination."
 Liquid bulk items in bulk containers can be protected from contamination by ensuring proper ventilation and by reviewing the condition of agitators routinely. Protection from contamination also means that CIP (clean-in-place) and product loops in distribution systems cannot be cross-connected. This is especially important in the beverage industry where most items are cleaned with CIP

systems. Also, review temperature controls on bulk holding tanks to ensure that the product can be and is adequately held to prevent growth of microorganisms.

The section on Manufacturing Operations requires that equipment and containers be maintained in an acceptable condition. Food manufacturing must be conducted to minimize contamination and growth of microorganisms, and therefore processors should monitor time, temperature, humidity, a_w, pH, pressure, and flow rate. Manufacturing operations such as freezing, dehydration, heat processing, acidification, and refrigeration should be monitored to ensure that mechanical breakdowns, time delays, temperature fluctuations, and other factors do not contribute to contamination of the food. Food shall be protected from metal by using sieves, traps, magnets, metal detectors, etc. Metal is not the only type of foreign material found in food. Your operations must ensure that controls are in place to adequately protect the consumer from objects such as stones, rubber, plastic, glass, etc. Manufacturing Operations continues on to describe various requirements of specific process steps: washing, peeling, trimming; heat blanching; batters and breaders; filling, assembling, packaging; dry products; and acid and acidified products.

Finally, the section on Warehousing requires that food be stored and transported under conditions that protect food against physical contamination and microbial contamination as well as against deterioration of the food and container.

Natural or unavoidable defects that occur at low levels that are not hazardous to health are covered within this section. Current defect action levels (pesticides, etc.) are not listed in this section, but reference is made to them. It is important to note that the mixing of food containing defects above the current defect action level with another lot of food is not permitted and renders the final food adulterated, regardless of the defect level of the final food.

BOTTLED WATER

Second on our path down Regulatory Lane are the Drinking Water regulations, 21 CFR Part 129. Again, comparable international standards exist in the Codex Alimentarius. "Processing and Bottling of Bottled Drinking Water" describes the general provisions and the conditions for buildings and facilities, equipment, and production and process controls. This section does not supplant 21 CFR Part 110 but is in addition to it. Specific requirements of this regulation include the separation of the bottling room from other plant operations or storage by tight walls, ceilings, and self-closing doors. Conveyor openings shall not exceed the size required to permit passage of containers. Washing and sanitizing of bottles shall be performed in an enclosed room. Cleaning and sanitizing solutions utilized by the plant shall be sampled and tested by the plant as often as necessary to ensure adequate performance in the cleaning and sanitizing operations. Records of these tests are required to be maintained, including a record of the intensity and duration of the agent's contact with the surface. Record retention for this industry is 2 years.

JUICE REGULATIONS

In February 2002, the new Juice HACCP regulations (21 CFR Part 120) were implemented. Within these regulations are specific requirements for sanitation: sanitation controls, monitoring, and records. Again, this regulation is in addition to Part 110.

Under the new regulations, each processor must have and implement Sanitation Standard Operating Procedures (SSOPs). SSOPs must address sanitation conditions and practices before, during, and after processing. Specific SSOPs must address:

1. Water — the safety of the water that comes into contact with the food including ice
2. Condition and cleanliness of food contact surfaces, including utensils and gloves
3. Prevention of cross contamination from unsanitary objects to food, packaging, and contact surfaces
4. Maintenance of handwashing, hand sanitizing, and toilet facilities
5. Protection of food packaging and contact surfaces from adulteration with lubricants, fuel, pesticides, cleaning compounds, sanitizing agents, condensates, and other chemical, physical, and biological contaminants
6. Proper labeling, storage, and use of toxic compounds
7. Control of employee health conditions that could result in microbiological contamination of food, packaging, and contact surfaces
8. Exclusion of pests

SSOPs must be monitored during processing with sufficient frequency to ensure conformance with 21 CFR Part 110. The regulations require that corrective actions be made in a timely manner. Your facility must ensure that corrections are taken, are timely, and are documented. Corrective actions only go so far, and when a problem is repetitive (twice is a repetition), preventive measures should also be evaluated and implemented. These preventive measures should also be documented.

Juice regulations require that SSOP records of monitoring and corrective actions shall be retained for 1 year for refrigerated (perishable) and 2 years for frozen or shelf stable products. Note that this is different from bottled water. Records shall include: the name of the processor, location, date, time, signature, and if appropriate the product/production code.

OTHER REGULATIONS

Many other regulations impact beverage companies. The Environmental Protection Agency (EPA) has regulations that cover the registration of chemicals (cleaners, sanitizers, etc.), and some requirements exist for containment barriers around large volumes of chemicals (for example, bulk storage of chemicals used in CIP processes). EPA also regulates processing water and wastewater. Already mentioned were FDA's parts 173 and 178 that govern boiler additives and sanitizers. FDA has other regulations that could impact your sanitation program.

State and local requirements should not be forgotten, particularly as they relate to liquids and solid waste. In a sanitation environment where large volumes of liquid and solid waste are created, these regulations are paramount to a good sanitation program. Monitoring the biological oxygen demand (BOD) of your wastewater, recycling chemical drums and barrels, etc. are all part of this process.

SUMMARY

What has been covered so far is not intended to be an all-encompassing survey of regulations. The goal of this discussion is to encourage processors to be regulation watchers. Read the regulations; outline your program against them. Read regulations that do not pertain to you — these give you a good idea of what other industry expectations are. For example, you might not be manufacturing bottled water, but your FDA investigator knows those regulations and hopes that you comply with the "higher standard." The SSOP requirements for juice are very similar to those for seafood. If you were aware of these regulations back in '96, you would not be surprised now. There is good information within the regulations that can be used as the base for your program.

WHAT TOOLS ARE USED BY THE BEVERAGE INDUSTRY TODAY IN FULFILLING ITS SANITATION PROGRAM NEEDS?

CLEANING AND SANITIZING

One major aspect of sanitation is cleaning. First, decide how you will clean a particular surface or piece of equipment. Wet or dry? Dry cleanups are used in bakeries and other food establishments where the use of water can be a food safety hazard. They focus on brush and vacuum cleaning, with care not to spread the soil. The beverage industry, however, primarily uses wet cleanups — rinse, wash, rinse, sanitize — through CIP and clean-out-of-place (COP) systems. Detergents, water, and sanitizers used in specific processes provide for effective cleaning. Selection of the detergent is important. Depending on the soil that should be cleaned, detergents can be acid or caustic. For most applications in the beverage industry, a high-pH chlorinated cleaning compound is used.

Water conditions must be included in the design of an effective program. If water is hard or soft, conditioning agents might need to be added to allow for the selected detergent to function at optimal levels. Water temperature must be appropriate for the type of chemical. Some detergents require a "soak" time before following up with mechanical action — brushing, scrubbing, etc. At one time, high pressure was used to rinse and clean surfaces; today, high pressure is not recommended where the spread of *Listeria* is a concern because high pressure loosens soil and can move it a considerable distance as an aerosol.

The selection of a sanitizing agent also is important. Chlorine, iodine, quaternary ammonia, and acid sanitizers are all used in the industry. The selection of the sanitizer is based on the target microorganisms and the cost. Be aware that sanitizers (chlorine and acid) can eat away at paint and equipment surfaces, causing a loss in equipment life. Again, a common application is quaternary ammonia, with the use of an acid-based sanitizer intermittently or for corrective action when a positive *Listeria* swab is found. Other types of chemicals can be used within the cleaning program — acid to remove scale buildup, fogging with acid, etc.

The most important key to cleaning in the beverage industry is the CIP system. Designed to replace mechanical (manual) scrubbing with laminar flow of the chemicals via a pump, this type of cleaning was made for the tanks and long, hard pipes common to the beverage industry. CIP systems, however, cannot be installed and "forgotten."

CIPs should be designed so the raw side of the operation is separate from the post-pasteurization operation. There must not be *any* opportunity for cross-contamination from raw to pasteurized juices or beverages.

CIPs should have current blueprints that should be verified every 12 to 18 months. It is rare that a system has been in place for a year without a little "tweaking." Maintenance activities and modifications to the system can also result in dead spots — areas that do not get cleaned through the CIP process. At least yearly, preferably quarterly or even monthly, the pressure or flow rate of the cleaning chemicals through the system should be measured. Since cleaning is based on the laminar flow, the flow rate is important. The flow rate should be within specification of the original design — too much pressure could be as detrimental to cleaning as too little. Many systems now are computerized where flow, chemical concentration (measured in conductivity), time, and temperature are all monitored and recorded. If your CIP is not so advanced, then either an external specialist in the field must be brought in or internal expertise will need to be developed to measure these important attributes of the CIP.

The COP system is a complementary task to the CIP. It allows for pipes, hoses, and parts not connected to the CIP to be cleaned through soaking and sometimes heat.

In summary, CIP sanitation has the same elements as other cleaning methods: a given chemical cleaner at a given concentration, used with a specified holding time and temperature, all complemented with mechanical action. Your CIP systems must be reviewed routinely to ensure that these elements are applied consistently and continuously. All CIP cleaning documentation should include at least these basic elements.

Procedures

Once the correct cleaning compound and sanitizer are selected, it is time to write the sanitation procedures. Procedures should be detailed work instructions for each piece of equipment and all areas of the facilities. How should the CIP system be set up? What buttons on the microprocessor should be selected? How should the mixing vat be cleaned? Scales? Dry ingredient storage racks? Refrigerators? All areas of the facility should have written procedures.

Procedures should follow standard quality management documentation: purpose, scope, activity, frequency, responsibility, records. Include special needs such as lock-out/tag-out and PPE (personal protective equipment).

Make sure the types of chemicals to be used for each task are specified, as well as the concentrations. Monitor and document concentrations of chemicals. Following weekly or monthly trends on chemical use can let you know if too much or too little is being used. And do not forget sanitizers. Even if you have a premeasured wall dispensing unit, the concentration of these chemicals must be checked on a frequent, routine basis, preferably daily. Premeasured dispensing systems can become clogged or the tips can slightly widen, allowing for a different volume to be dispensed. It does not take much change for the concentration of the chemical to go from 200 ppm to 100 or 400 ppm, when we are talking about such a large dilution factor. Once the frequencies of all tasks are determined, a Master Sanitation Schedule can be created. Finally, monitor and verify the effectiveness of the program.

OTHER SANITATION ELEMENTS

In addition to cleaning and sanitizing equipment and areas of the facility, a good sanitation program includes pest control. Whether pest control is performed internally or externally, the personnel must be certified applicators, appropriately licensed in the state or province to handle, mix, and apply pesticides. A detailed pest control program should be written. (*Note:* the contract with a pest control company is not a detailed program.) Include all types of pests and how those pests are managed. Insects, birds, and rodents should be included. Proper documentation — insurance certificates, material safety data sheets (MSDSs) and pesticide labels, applications, trends, violations of integrated pest management principles (attractants, exclusions), should be available. Supporting programs — locker cleaning program, break room cleanliness, building integrity, grounds, dumpsters, etc. — should be integrated into the pest control program.

Next, check the maintenance department. The preventive maintenance aspects of the equipment directly impact the cleanliness of the facility. Room air (HVAC) should be controlled to minimize possible contaminants. Compressed air that is directly impinged onto product or product contact surfaces must also be of appropriate bacterial quality. The maintenance staff must have a documented routine program that ensures that these filters are reviewed and maintained at the appropriate level. Some plants move dry ingredients to the blending tanks using what we will call pneumatic air. The source of this air must also be appropriately controlled and maintained.

As mentioned earlier, steam that may come into contact with a food product must be of appropriate sanitary quality. This means that any time you have only a single barrier (stainless steel wall) between the steam and the product, the steam must be assured to be of a culinary quality. All boiler water treatments must meet the requirements of 21 CFR 173. This holds true for recirculated water used to cool or heat products, too. These waters must be tested and treated to prevent contamination of the food product.

Lubrication must be food grade in any location where the lubricant could contaminate product. If the lubricant is not food grade, the plant must have appropriate controls in place to properly dispose of the food if it may have become contaminated. Maintenance programs must be in place to ensure that these items are routine and are documented. Many plants have a preventive maintenance program for food safety to address these critical needs.

Finally, the sanitation program cannot ignore the special needs of the beverage industry. The production of most types of beverages involves filters of some type. These must be checked regularly to be sure that the filters, and any materials that they might have trapped, are not acting as sources of contamination. Borrowing from the bottled water regulations, a separate room for filling is a superior sanitation enhancement. Consider a "high hygiene" area where forklift and foot traffic are limited, special air filtration (HEPA filters) is used, and personal hygiene has a higher level, such as the use of jumpsuits and head coverings (not just hairnets). Air and water are also critical elements in beverage sanitation. Any time water is used on equipment or bottles and caps after the final microbial reduction step, a heightened awareness is needed. Transmission of protozoa through water must be considered in the sanitation environment. The plant should have a water quality profile — microbial tests of water throughout points of use in the plant — and keep this current on a yearly basis. The same is true for air — any air blowing into the filling room, bottles, etc. must be clean. Air quality profiles (usually yeast and mold) are also valuable tools.

Master Sanitation Schedule

Once the frequency of sanitation tasks has been established, creation of the Master Sanitation Schedule can ensure that each task is completed and monitored. Creating a Master Sanitation Schedule can be done in many ways. Figure 8.2 gives one example.

The Master Sanitation Schedule is a calendar of sanitation tasks. Record when tasks are completed on the schedule. Make sure to note when tasks are not completed and why. Modify the frequencies as needed. Perhaps a task needs to be completed more frequently in summer than in winter. There is no rule that says that a task has to be completed "quarterly." Perhaps every month for six months and then once in the latter half of the year could be the desired frequency. Setting frequencies requires an understanding of the history of the equipment and facility. Be prepared to modify and update the schedule as needed.

TASK	TIME.	1	2	3	4	5	6	7	8	9	10	11	12	13	14	15	16	17	18	19	20	21	22	23	24	25	26	27	28	29	30	31
Chipper	Daily																															
Table	Daily																															
Trash Barrels	Daily																															
Floor Mats	Daily																															
Hopper	Daily																															
Extractor	Daily																															
Drains	Weekly																															
Walls	Weekly																															
Overhead Pipes	Weekly																															
Outside of Tanks	Weekly																															
Raw Bins	Weekly																															
Dumpster	Weekly																															
Ceiling/Light Covers	Monthly																															
Chem. Storage	Monthly																															
RM Cooler	Monthly																															
Air ducts	Quarter																															
Ware. Racks	Quarter																															

FIGURE 8.2 Example Master Sanitation Schedule.

And do not forget to document. Most companies use a large master scheduler or calendar. Some companies use their preventive maintenance software for sanitation tasks. Whatever system allows for the easiest, most compliant recordkeeping process should be used. And do not be afraid of logbooks. Some sanitation managers keep a bound notebook with special requests, trouble spots, or other information (for example, the sprinkler system was checked). Such a "diary" has been very useful when looking for the root cause of a problem that surfaces months later.

HOW IS THE EFFICACY OF THE PROGRAM VERIFIED?

The most frequent type of verification of sanitation activities is daily reviews. These reviews should be both preoperational and operational. The checks should be designed to ensure that all plant areas and equipment are clean and ready for use. Reviews should be conducted by a person or department independent from Sanitation. Sanitation cleans and sanitizes; other departments in the operation verify the effectiveness of that work. Think of the Internal Revenue Service; it reviews tax returns prepared by the taxpayer. It would be nice to audit your own tax return, but that is not appropriate. Some companies have quality assurance (QA) conduct these reviews, while others use production supervisors, with QA verification once or twice per shift.

Also, routine full-plant audits should be conducted. The norm for these is monthly audits conducted by the QA manager. However, more creative companies are dividing the plant into areas and having a management team member audit each area once per month. The team members rotate their assignments. Others use a management team inspection where someone from each department participates in the audit. But do not forget — all of these are recorded and, most important, corrective actions are generated and documented. If a maintenance activity is needed to fix a discrepancy found on a routine monthly inspection, then document the work order, track it, and note its completion as part of the corrective action plan. Corrective actions also should be documented for daily reviews. Many companies do not record corrective actions because "it was fixed right away." This type of philosophy does not allow for trends to be monitored and continuous improvement cycles instituted. Corrective actions are a documented path to follow to determine where preventive measures are needed. Also, do not forget verification of the CIP system. When blueprints are reviewed or the pressure/flow rate is monitored, document these verifications.

Another verification tool is the use of bacterial or adenosine triphosphate (ATP) swabs to determine the efficacy of cleaning. Bacterial swabbing measures indicator organisms that affect product safety and quality. Turn-around time on these bacterial tests can be 24 hours or more, which means Production most likely has used the equipment since the test was taken. An unsatisfactory reading indicates that the cleaning program for this area needs to be enhanced and can be a valuable tool for the sanitation crew. The ATP method uses a luminescence process that gives an instant (within seconds) measurement of the amount of ATP on the surface. As ATP is a chemical within cells, it is a measurement of "dirt" or organic matter on the surface, not necessarily bacteria. Its benefit is that it allows for immediate response to an unsatisfactory reading. In either case, there are a variety of ATP tests/types of bacterial swabbing methods that can be used. As long as the testing allows for the efficacy of the procedures to be verified and continuous improvement to be made, either will be adequate.

Another good verification tool for sanitation programs is the use of environmental swabbing or sponging to determine whether pathogenic microorganisms, usually *Listeria* sp., are present. An environmental program focuses not on the contact surfaces but on the noncontact surfaces and the environment. This determines whether a pathogen has found a niche from which it can survive, grow, and become a potential problem. By locating niche areas, processors can focus on these areas to prevent the "explosion" of the organism onto product contact surfaces.

It is important to test the environment where you expect problems: under door frames (particularly those of coolers); drains and the backside of drain covers; wheels on trolleys; barrel stands; lift tracks for hoisting equipment; condensing units; etc. Create a map of locations that you want to test and select a few every month until you develop your pattern and history. Then focused sampling of "hot spots" and random sampling of other areas can continue. Any environmental pathogen testing program should have a corrective action program if a positive sample is found. The purpose of environmental testing is to confirm that your routine sanitation program is capable of removing pathogens and keeping them out of the facility. When a positive result is noted, corrective actions should include:

1. Immediate recleaning of the area to prove that it can be cleaned and pathogen free
2. Routine sampling of the area to prove that it can be maintained clean and pathogen free
3. Review of the routine sanitation procedures for that area

CASE STUDIES FOR BEVERAGES

The critical sanitation issue for beverages is post-pasteurization contamination. This includes contamination of the pasteurized product as well as any ingredients added to the pasteurized liquid (vitamins, purees pasteurized in another facility, etc.). If ingredients are added post pasteurization, these ingredients must be controlled within a vendor selection, certificate of analysis (COA), and receiving program. Even the lab that performs the analysis should be part of the control program.

Outside of ingredients and supplier controls, what about internal processes and controls? Let us look at some of these post-process contamination case studies. The first area is pumps. In a beverage company not too long ago, a pump was being dismantled because of a maintenance issue. As the cover was removed, cockroaches streamed out of the pump. Another company recently tracked a sporadic mold problem to a pump that was not cleaned properly. Product pumps and CIP pumps can house problems if they are not properly cleaned. Review of the CIP loops to ensure that each item in the process flow is being cleaned is a vital tool in preventing problems such as these. When purchasing pumps (and there are multiple types — positive displacement, diaphragm, etc.), learn the proper cleaning techniques first. Take the time for a full cleaning of the pump each and every day it is used.

The second sanitation case study involves air filters. An air conditioning unit exhaust was placed above a fill tank in one company. In trying to determine the source of the yeast growing in the beverage (and these were some hearty yeast cells — living even after 200°F heat treatment), an air duct was discovered that resembled the surface of the moon — or worse, a mushroom farm — because of the amount of yeast growth on it. The intensity of the yeast and dirt substrate growing in the duct had, in some places, almost filled the diameter of 12 to 16 in. Keep air ducts and filters clean. These should be on the Master Sanitation Schedule. Try not to place exhausts over product fill lines or open containers.

In another case study, the source of spoilage organisms and reduced shelf life was a dead end in the CIP system that had not been discovered because the system had not been verified since 1992, when it was installed. We mentioned earlier that CIP loops should be verified on a routine frequency of 12 to 18 months. Each situation is unique, and the plant staff must decide on the appropriate frequency of inspections. It is important that not just CIP loops but also product loops be verified to ensure no cross connection opportunities with CIP liquids or pasteurized products. A dead end on a line can be as short as 1 inch if CIP solution flow (mechanical action) is insufficient to clean it. All modifications to the CIP system should be reviewed by competent staff before the changes are approved for installation and use.

Last, rough welding on the inside of pipes caused another company to have sporadic shelf life problems. Yeast and mold growth would occur in the finished product while in retail cases, causing spoilage before the shelf life expiration. Because the shelf life issues were sporadic and occurred over the entire range of products (different flavors, different sizes of packaging), tracking the source was very difficult. Rough welds were found in some transfer pipes. Product buildup at the welds would occur, and at sporadic intervals the buildup would "let go" and contaminate the product flowing over the weld. The 3-A program requirements for welds on any beverage pipes should be reviewed with the welders at the installation stage to ensure that contractors and in-house workers have the appropriate skills and that their work is reviewed by competent staff.

CONCLUSION

Sanitation has been considered the poor stepchild to HACCP in recent years. Training, education, and resources have been focused on HACCP. Sanitation is as important as HACCP, however. Consider the two programs as train tracks. The food safety train will derail if both are not intact.

It is a good idea to use the concept and principles of HACCP when designing your sanitation program. While sanitation is a separate and distinct program, the principles — hazard analysis, critical control points, critical limits, monitoring, corrective action, recordkeeping, and verification — have application and merit in designing a sound sanitation program.

A good sanitation program relies on trained, educated, skillful, and knowledgeable sanitation staff. One of the difficulties in establishing and maintaining a sound sanitation program is the lack of good education and training for sanitation managers. No certification program or continuing education program for sanitation personnel exists. As a base to a company's food safety program, sanitation is essential, and employees in this area would benefit from the support and resources of a continuing education program.

In designing and maintaining the sanitation program, keep a few key points in mind:

1. Be a regulatory watcher.
2. A beverage sanitation program has numerous needs, but the CIP and filling area are two key essential elements (as are air and water).
3. Document your sanitation program.
4. Verify your sanitation program.

And, above all, keep producing those enjoyable, high quality beverages we have come to love!

9 Juice Processing — The Organic Alternative

Susan Ten Eyck

CONTENTS

Introduction 193
History of the Organic Movement 194
Market Demand for Organic Products 194
Organic Processing and Regulations 195
Sanitation 197
Processing of an Organic Product 198
Federal Regulations 200
Certifiers and the Certification Process 201
The Future of Organic Products 203

INTRODUCTION

What is organic? To define this term, we need to understand the concept behind organic production and organic processing.

There are three degrees of commitment to growing and processing food products in an organic manner. The deepest is the philosophical commitment, which borders on a reverential, almost religious dedication. Those who subscribe to the organic philosophy consciously and faithfully produce and consume only food and fiber that is produced in an organic manner without synthetic fertilizers or synthetic pesticides and grown in harmony with nature. Processed foods and fiber are produced with minimal processing and processing aids.

The second degree of commitment is to organic principles. This involves growing food in a sustainable manner by replenishing and maintaining soil fertility with crop rotation, cover crops, composted fertilizer, and natural minerals, as needed. The organic farmer believes the soil is living and needs to be nourished to produce healthy food crops. Biodiversity minimizes pests, weeds, and disease and reduces soil erosion. Beneficial insects are used in place of pesticides, with hedgerows planted to provide habitats for these beneficial insects. All growing is to harmonize with nature.

The third degree of commitment is to the practices of growing and processing in an organic manner. These practices are being conducted and recognized around the world, some out of choice and some due to necessity. Organic growing practices within the United States are on land free from prohibited substances for three years prior to growing an organic crop. No sewer sludge is applied to the land, and the use of synthetic fertilizers and pesticides is not permitted. Cessation of chemical application to the land does not constitute organic practices. This is sometimes called "organic by neglect."

HISTORY OF THE ORGANIC MOVEMENT

The organic movement, as we know it today, began with Sir Albert Howard, who combined scientific training with the study of the traditional composting methods of India and China. In 1940, he advocated that Britain preserve what he called the cycle of life by adopting sustainable agriculture using urban food waste and sewage. His published articles were reprinted in the United States and were a great influence on Jerome Rodale. It was Rodale who coined the term "organic." In 1942, Rodale published *Organic Gardening and Farming*, which greatly influenced the American small farmer. Then in 1946, Lady Eve Balfour established the Soil Association in Britain. The main purpose of this organization was to unite those people working toward a more complete understanding of the vital relationship among plant, animal, and man. From 1950 through 1962, scientific farming with an emphasis on chemical fertilizers and chemical pesticides was becoming the hallmark of agriculture. During this time, a small contingent of farmers shunned the use of chemicals in agriculture and followed the guidelines established by Rodale. In the United States, these farmers were considered members of the counterculture, nonconformists, and even hippies. They were growing crops in this manner solely out of personal commitment to the land.

The wake-up call about the heavy dependency on chemicals in agriculture occurred in 1962, with the publication of Rachel Carson's *Silent Spring*. This book described the destruction of wildlife and threat to human health from widespread use of chemicals in commercial farming. *Silent Spring* drew attention to the harm to avian life resulting from the use of pesticides, especially DDT. As a result of this book, the use of DDT was banned in many countries. This is the background that gave birth to organic food production as we know it today.

MARKET DEMAND FOR ORGANIC PRODUCTS

There is a growing market demand for organically grown and produced food products. This market demand is driven by the desire to maintain or improve health, and people believe that organic growing practices are more environmentally friendly. Typical organic consumers are educated females in their mid-thirties to mid-forties with children at home. The organic market has been growing and continues to grow at the rate of 20 percent a year. Consumers have perceived a value added benefit to purchasing organic foods. With increased consumer demand, the cost of organic production has decreased due to economies of scale. As consumer demand increased, the major retail markets have added the organic category to their everyday variety of food. Organic foods are no longer relegated to small natural food or health food stores. Currently, organic food holds 2% of the food market share. The industry predicts that organic food will have 5% of the food market share by 2006. Thus, rapid growth is expected in the next few years.

Many of major food companies are already processing organic foods such as General Mills with cereals and flour. General Mills has also recently acquired the Small Planet Foods Co. Kellogg's, Dole, and other national grower/shippers such as Duda and Driscoll have also entered the organic market. Why have these national food processors entered the organic food processing business? They see the market demand and potential for market growth in organic foods. Dramatic market changes will occur in the next couple of years.

What does it take to enter the organic food processing market? It takes desire. The organic industry is now a regulated industry; it is regulated by the United States Department of Agriculture (USDA) and further monitored by the Food and Drug Administration (FDA) and the Environmental Protection Agency (EPA). As such, every company in every state will follow the same regulations. And any food or product labeled "organic" will need to comply with the USDA organic regulations. If you are ready to produce an organic beverage, the first step should be to contact a major organic certifier. Certifiers are accredited by the USDA to certify organic products. USDA announced the first round of organic certifiers in 2002.

ORGANIC PROCESSING AND REGULATIONS

As an introduction, let us start with a brief overview of organic processing. Organic processed foods are safe. Organic processing must comply with all FDA regulations and all state and local health regulations. While organic products are grown without pesticides and herbicides, the organic industry does not claim that products are totally free from pesticides and herbicides. These synthetic chemicals are in the atmosphere because of drift from nonorganic applications. Within processing facilities, during organic production, exposure to pesticides is prevented. Organic processing is philosophically minimum processing with minimal nonorganic ingredients. Some synthetic ingredients and processing aids are permitted when there is no organic alternative. For food safety, citric acid is permitted as a pH regulator; for functional requirements, baking powder is permitted. Organic food products are not marketed as healthier than nonorganic versions because insufficient scientific data are available to validate this claim. Yet some consumers perceive organic products to be healthier. Some organic consumers, because of serious food allergies, find organic products more agreeable in their diets.

The rules and regulations that control organic growing and processing are published in the Federal Register, Part IV, Department of Agriculture, Agricultural Marketing Service, 7 CFR Part 205 National Organic Program; Final Rule. Under this federal regulation, any producer growing products to be labeled organic and any handling operation processing food that will be labeled as organic must be certified by a USDA-accredited certifier. Organic growers, now called the producers, must be certified to verify that the production system in place is designed to optimize soil biological activity, maintain long-term soil fertility, minimize soil erosion, and maintain and enhance genetic and biological diversity in production. Handling operations that further process the organic agricultural products must be certified to verify that they implement organic good manufacturing and handling practices in order to maintain organic integrity of the products. An intermediary operation between the producer and the handling operation that takes possession of the organic product, such as a cold storage operation, must also be certified. Truckers and distributors of further processed packaged organic foods do not require certification. Likewise, retail operations do not need to be certified unless they are further processing organic products. The purpose of the certification process of the handling operations is to ensure that the practices minimize environmental degradation and minimize the consumption of nonrenewable resources. They further verify there is no commingling with nonorganic ingredients either prior to processing or during the processing operation and that the product and packaging material do not come in contact with prohibited materials.

What is the composition of an organic juice or beverage? The main ingredients must be organically grown agricultural products — fruit or vegetables. When the beverage is labeled as "100% Organic" juice [7 CFR 205.301(a)], the entire product must be organic with no synthetic ingredients. Not even the synthetic ingredients approved for use in 7 CFR 205.605 and 205.606 are permitted in this category of organic products. Processing aids must also be 100% organic; for example, organic rice hulls as a filter aid and organic lemon juice in place of citric acid for pH adjustment. Products labeled "Organic" [7 CFR 205.301(b)] must consist of a minimum of 95% organic agricultural ingredients (excluding water and salt), and the remaining ingredients must be nonagricultural substances, non-organically produced agricultural products on the National List of Allowed and Prohibited Substances (7 CSR 205.605 and 205.606), or ingredients that are commercially unavailable in organic form. Individual USDA-accredited certifiers will determine the "commercial availability" of the specific ingredient based on documentation supplied by the handling operation. Neither of these label categories, 100% Organic or Organic, may contain any ingredients that were grown with sewer sludge or handled with ionizing radiation as described in FDA 21 CFR 179.26 or contain any ingredient or processing aid that was genetically modified. A product may be labeled "Made with (*ingredient category*)" when 70 to 95% of the beverage is produced from an organically grown fruit or vegetable. In this category, only the organic ingredients must comply with the restrictions mentioned above. There are no restrictions on the remaining ingredients. Wine will be labeled "Made with Organic Grapes." Sulfur dioxide, on the National List of Approved and Prohibited Substances, is annotated for use in wine labeled Made with Organic Grapes, at a total concentration no more than 100 ppm. Any product that contains less than 70% organically grown ingredients may not be labeled organic. Reference to the organic ingredient may be made on the ingredient panel only.

A vitamin-fortified or nutraceutical organic beverage or juice is not possible under the National Organic Program. Nutritional vitamins and minerals must be in accordance with 21 CFR 104.20. There are provisions in the federal regulation to petition for additional ingredients and processing aids to be included. The petitioner needs to document the need for the inclusion. Recommendations will be made to the National Organic Standards Board, which will request inclusion in the Federal Regulation.

SANITATION

Sanitation is a critical area in the processing of organic products, especially if the processing facility handles both nonorganic and organic products.

All normal sanitation procedures must be in place; in addition, thorough cleaning of all processing equipment must take place prior to the running of organic product to prevent commingling with nonorganic product. Care must be taken to eliminate all sanitizer residues on equipment that will come in contact with organic product. A thorough detergent wash followed by a thorough rinse is required. Frequent testing is required to verify the total elimination of sanitizer, and an extra rinse will be used. Sanitizers permitted in preparation for organic processing are restricted to chlorine materials, hydrogen peroxide, peracetic acid, phosphoric acid, and sodium hydroxide. Detailed procedures used in equipment cleaning and sanitizing must be provided to the certifier in the Organic Handling Plan. In a facility that processes both organic and nonorganic products, it is necessary to maintain sanitation logs documenting a thorough cleaning prior to the organic processing. Any piece of equipment that comes in contact with organic material must be cleaned, including scoops, knives, bins, hoses, kettles, and filling machines. Should there be a piece of equipment that cannot be cleaned with detergent and water, it may be possible to use a purge with organic product to clean the system. When a purge is used for cleaning purposes, the quantity of product must be documented, along with the duration of the purge and the disposition of the purge material. This purge documentation needs to be signed by the supervisor on duty at the time of the action. The use of purges and the quantity of material used must meet with the approval of the accredited certifier.

PROCESSING OF AN ORGANIC PRODUCT

The organic production process starts with receiving the raw materials and packaging materials. Receiving logs document incoming organic raw materials and include the material lot number in the log. Upon receipt, receiving personnel must verify that the bill of lading states that the ingredient is organic. If the raw material is not identified as organic on the incoming bill of lading, the ingredient may not be used as organic; the processor must either refuse delivery or use the ingredients in nonorganic products. Incoming raw materials and packaging materials need to be stored in a designated area for organic materials that is free from the possibility of contamination with prohibited insecticides and fumigants and also free from the possibility of cross-contamination.

In an operation that is not dedicated to organic processing, the organic products are usually processed first in the morning. After production supervision verifies that the process equipment is thoroughly cleaned, production can proceed. Pasteurization is an acceptable organic process. The current restriction on heating is the prevention of product contact with volatile amines in boiler chemicals. Steam injection prior to capping would be an area in which this could be a problem. It is easily solved by shutting off the boiler chemicals prior to the organic run, purging the steam chemicals, completing the run, and then returning to the use of boiler chemicals for nonorganic processing. National Organic Standards Board members are discussing the possible inclusion of the current technology into organic processing. No decisions have been made at the time of this writing. The packaged product needs to be stored in a designated area away from nonorganic product to prevent accidentally shipping nonorganic product under organic documentation. Each step in production needs to be documented with lot numbers of raw materials, time of process, quantity of product produced, quantity packaged, and lot number of finished product. An audit trail of organic product movement, from incoming raw material through to the finished product shipment to the customer, must be clearly maintained. All documents relating to organic production must be clearly marked as Organic, including logs, production records, shipping documents, and invoices. This clear audit trail needs to be complete enough to immediately trace any product suspected of contamination from the point of origin (the grower) to the processor's customer (the consumer).

During the processing of an organic product, the main criterion is to maintain the organic integrity of the organic ingredients. This maintenance of integrity starts with the organic processing scheduling. When changing from nonorganic processing to organic processing, a thorough cleanup is needed to prevent any possibility of commingling of residual nonorganic materials remaining in the processing equipment. Nonorganic ingredients need to be removed from the processing area. Should there be some incidental commingling with nonorganic product, the organic product ceases to be organic and must be labeled nonorganic. Most organic processors will run the organic products at the start of the day, when all equipment is clean and there is no opportunity for commingling of organic ingredients with nonorganic ingredients. If the processor must switch to organic processing in mid-day, there is usually a lengthy down time due to the sanitation procedures that must be completed to prevent any contamination from nonorganic products. When production is from raw organic ingredients, often the organic processing can be scheduled to run for several days.

Organic ingredients and finished products need to be protected from contamination by prohibited pest control products. All pest control measures must follow Good Manufacturing Practices (GMPs), with the exception of facility fumigation. If the processing area is fumigated, all organic ingredients and packaging material must be removed. The reentry time for organic ingredients and packaging material is 1/2 times the label reentry time. Equipment will need to be thoroughly cleaned of any possible prohibited material residue prior to organic processing. The Federal Regulations in 7 CFR 205.271 are not as restrictive about the use of fumigants as private certifiers have been. However, a thorough cleanup prior to organic processing is still necessary. The steps to pest control for organic processing, under the Federal Regulations, now state that the following preventive measurers must be in place: prevention, then control with mechanical or physical controls, followed by the use of nonsynthetic repellents and lures or synthetic substances that are on the National Materials list (7 CFR 205.605). Only if the previous steps are not effective in eradicating the infestation can synthetic measures, such as fumigation, be used. The processor must not rely on synthetic measures. After the infestation has been eradicated, the processor must return to the less invasive measures. Only if the processor can thoroughly document the necessity of synthetic measures will the certifier grant the processor permission to use such materials. At all times, organic packaging material must be protected from contamination with pest control materials. Care must be taken to prevent contamination from pest control products or fumigants during finished product storage and product transportation.

We have talked about the importance of the organic audit trail — the paper trail that is an integral part of the documentation for organic processing. This audit trail allows for total traceability of every ingredient in the organic product. An organic ingredient in a product purchased at retail can be tracked back to the organic field in which the ingredient was grown by means of the code dates on the retail product and by lot numbers. Accomplishing this audit trail or paper trail begins with the documentation of incoming organic raw materials. All incoming bills of lading must state that the product is organic. Copies of the organic certificates need to be available for the receiving department to verify that the incoming ingredient is indeed a certified organic ingredient.

All documents relating to organic processing must be clearly marked "Organic" to facilitate document tracking. Batching sheets or processing documents need to have the lot number of each ingredient, and the processing documents need to refer to the code date of the packaged product. The outgoing shipping records must be noted as "Organic" and contain the code date or the lot number of the product being shipped. The verification of this audit trail will be a major part of the organic inspector's annual visit. This audit trail will be used to verify that the organic ingredients bought are sufficient to produce the products shipped out.

FEDERAL REGULATIONS

While the Federal Regulations are now law, there was an 18-month phase-in period to allow everyone to become fully compliant. On October 21, 2002, all organic products were required to be compliant with the Federal Regulations. Any product labeled organic must be certified by a USDA-accredited certifier. All product labeled as 100% Organic, Organic, and "Made with _____" must identify the certifier. The statement "Certified organic by [the name of the certifier]" must be on the information panel below the information identifying the handler (producer) or distributor.

Multi-ingredient products labeled as 100% Organic must be all organic, and any processing aids used in the handling process must be organic. A product labeled Organic must contain 95% raw or processed organic agricultural products excluding water and salt. The remaining 5% of ingredients are nonagricultural or nonorganic ingredients that appear on the National

List of Allowed and Prohibited Substances or must be ingredients that are commercially unavailable in organic form, with full documentation. Products that contain 70 to 95% organic ingredients, excluding water and salt, are labeled "Made with Organic _____." A maximum of three ingredients may be listed, e.g., "Made with Organic Tomatoes, Onions, and Basil" or "Made with Organic Flour, Honey, and Oils." For a product labeled as "Made with Organic_____," the percentage of organic ingredients must be declared. The remaining ingredients are restricted to the extent that they may not be derived from genetically modified organisms, grown with sewer sludge, or processed using ionizing radiation. This "Made with Organic _____ " label may not show the USDA Organic seal, but the certifier seal is permitted.

Multi-ingredient products that contain less than 79% organic ingredients may not identify the product as organic or made with organic ingredients. The identification of the organic ingredient is permitted only in the ingredient statement with the percent of organic ingredients noted. There are no restrictions on the other ingredients in the product. Neither the USDA seal nor the certifier seal may be displayed on the label.

CERTIFIERS AND THE CERTIFICATION PROCESS

All these regulations are well and good but how are consumers to know that the products they are purchasing are truly organic? This is accomplished by Federal Regulations. The USDA's National Organic Program has control over the term "Organic." Anyone selling a product labeled "Organic" must be certified by an organic certifying agent accredited by USDA. Since October 21, 2002, any food product labeled as organic or ingredient identified as organic must be certified by a USDA-accredited organic certifier. These accredited organic certifiers can be private nonprofit or for-profit organizations or state certification programs. To become accredited, certifiers had to apply to USDA by October 21, 2001, and show they could operate in compliance with the certifier requirements stated in 7 CFR 205. The applications were reviewed, and the certification operations were audited by members of USDA for compliance with the regulations. Those certifiers that could demonstrate compliance were announced on April 21, 2002. At this time, those operations that were certified by the announced accredited certifiers were considered to be in full compliance with the Federal Organic Regulations.

The cost of organic certification to the producing and handling operations will vary among the certifiers; however the proposed fees charged by the certifying agent must be reasonable, and these fee schedules must be submitted to the National Organic Program administrator. Generally, the cost of certification depends on the size and complexity of the operation.

Costs, including the organic inspection of the operation, can vary from around $1000 to over $10,000 a year.

The certification process begins with the completion of the organic handling plan, which details how the integrity of the organic product will be maintained during processing and packaging. The organic handling plan must document how the organic product can be tracked from the field in which it was grown to the processed product by means of lot numbers and documentation. This is easily accomplished with a good recall plan. The application for certification is sent to the accredited certifier of choice with the required fees. When the application is reviewed for completeness pursuant to the Federal Regulations, an on-site inspection is arranged. The inspector writes a comprehensive report of inspection observations and notes any noncompliance issues. This report is reviewed by the certification agency. If the certifier determines that all procedures and activities stated in the organic handling plan, submitted with the application of the applicant, are being followed and no major noncompliance issues are present, certification is awarded. The operation can be certified with minor nonconformance issues. Within a reasonable amount of time, established by the certifier, the minor nonconformance issues must be corrected, and written corrective action measures must be submitted to the certifier. Under the National Organic Program, certification is good until revoked providing the operation is inspected annually. These inspections are similar to International Standards Organization (ISO) or American Institute of Baking (AIB) inspections.

Organic inspectors are trained in organic processing procedures and organic critical control points. Most processor/handler inspectors have completed the Independent Organic Inspectors Association (IOIA) training course, and many have completed advanced processor inspection courses or related auditing courses.

The inspector will come into the operation during an organic run to verify that the procedures specified in the application are actually being followed. At the discretion of the certifier, water samples and product samples can be collected and tested for prohibited substances. During the inspection, documents relating to the organic processing or organic ingredients are audited. Verification of the audit trail and audit of sufficient ingredients to produce the organic inventory are critical parts of the inspection. This step is a deterrent to fraud. The inspector will also interview key personnel in the process and documentation portion of the operation, verifying that the personnel are knowledgeable of the organic processing and documenting standards. The inspector will then conduct an exit interview with the key personnel outlining the nonconformance areas observed. The inspectors cannot advise the operation on measures to take to become compliant. A copy of the inspector's report is sent, by the certifying agent, to the inspected operation along with the certification decision and the timeline by which the minor noncompliance issues must be addressed.

The National Organic Program, USDA Agricultural Marketing Service (AMS), is a source for general information. The Organic Trade Association has current information on available certifiers on its Web site at www.ota.com. Many states have organic certification programs, in particular Washington, Texas, Colorado, Nevada, Kentucky, and Pennsylvania. These programs can be contacted through the state's Department of Agriculture.

THE FUTURE OF ORGANIC PRODUCTS

The future seems bright for organic processed products. Organic processing is compatible with nonorganic processing. Increased demand for organic products has been fueled by an increase in generic organic advertising and marketing. Predictions are confidently voiced that by 2005, organic products will capture 5% of the retail market. We have come full circle and are poised on the verge of a new era of working in harmony with nature for the betterment of humankind and Earth. The New Old is here now.

10 Active Packaging for Beverages

Paul L. Dawson

CONTENTS

Introduction 205
Food Labeling 207
Oxygen Scavengers/Antioxidants 207
Antimicrobial Polymers 210
Bio-Based Materials for Packaging 215
Taint Removers 215
Conclusion 216
References 216

INTRODUCTION

Active packaging can be defined as "packaging that performs a role other than an inert barrier to the outside environment" (Rooney, 1995a). Some crude examples of active packaging cited by Rooney (1995a) include wine skins that collapse with removal of the wine to maintain a minimal headspace in the package and tin-lined cans to prevent corrosion of iron in cans. The traditional wine bottle has several "active" components including colored glass, which prevents light damage; the cork, which is kept damp by storing the bottle horizontally to improve the oxygen barrier; and the tin layer, which prevents contact between lead and the wine.

More advanced types of active packaging, such as oxygen scavengers, were produced as early as 1938 in Finland. Different active packaging types have been produced in response to specific needs of the product. "Smart" films have been used in horticulture products longer than in other products to maintain an ideal gas atmosphere for slow respiration. These smart films now include oxygen scavengers to create a low oxygen environment, ethylene scavengers to keep this plant-ripening hormone at low levels, and carbon dioxide releasers that slow plant tissue respiration. Active packaging has also been applied to other foods such as high a_w bakery products, for which ethanol-releasing sachets can be used to suppress mold growth. Microwave susceptors actively heat and alter products for consumption; examples include popcorn and portions of prepared meals.

A specific active package type is not normally applied across a broad spectrum of food products. Rather, it is applied to a specific niche to extend the quality or safety of that product. One such example of a specific niche is self-heating cans of sake. Aluminum cans are heated by the controlled mixing of lime and water. Wagner (1989) reported that 30 million such cans were produced in 1988. This process was also applied to coffee containers and lunchboxes. Self-cooling cans have also been developed, using the reaction between ammonium nitrate and chloride. A rather large niche is oxygen-scavenging closures for beverages such as beer.

Brody (2001a), in reporting on international food packaging meetings, differentiated between active and intelligent packaging, defining active packaging as systems that sensed environmental changes and responded by changing properties. He further differentiated that intelligent packaging measures a component and signals the result. Examples given of active packaging include oxygen absorbers, antimicrobials, and controllers of moisture, odor and gases. Intelligent packaging includes antitheft indicators, locating devices, and time–temperature sensors. An example of a unique use of time–temperature sensors is indicators on special containers of Hungry Jack Pancake Syrup to indicate the optimum serving temperature during microwave heating. The definition of active packaging may be too narrow in that it implies that an environmental change must occur for the package response to occur. Antimicrobial and antioxidant packaging will release active components to the food without an environmental change. Using a broader definition, active packaging acts on the food product to maintain quality or change the food for consumption.

Most active packaging applications are used to maintain the quality of the product. The quality factors that deteriorate most quickly in beverages are related to oxidation and microbial growth. Oxidation can alter color, flavor, and nutritional value, while microbial growth can affect these factors as well as safety. Since oxidation requires oxygen, a common method to slow this reaction is exclusion and removal of oxygen from the package. Oxygen scavengers or absorbers can be included in packaging systems as sachets, as closures (crowns), and in polymers. Iron-based scavengers have dominated the scavenger market; however, other systems have been introduced that use ascorbic acid in combination with other organic and inorganic compounds. Antimicrobial films have not had the same widespread application as oxygen scavengers in beverages. The most discussed antimicrobial packages have been those containing silver ions or salts dispersed in zeolite. These were first introduced in Japan. Silver has been incorporated into polymer coatings, which are used to coat metal surfaces, by Agion. These products are marketed by AK Steel. The use of oxygen scavengers and antimicrobials will be discussed in more detail in later sections of this chapter. Other topics covered will include food labeling regulations, antioxidants, bio-based packaging and taint removers.

FOOD LABELING

Active packaging systems may sometimes require that a component migrate from the package to the food. This has relevance to food package labeling in that the food contact surface of a package must be proven to be safe. That is, any compound that migrates from the package into or onto the food is considered a food additive. Food additive requirements include that the additive:

1. Must be safe at the intended use level
2. Must perform a function
3. Must not mask a property
4. Must not reduce nutritional value
5. Must not replace a Good Manufacturing Practice (GMP)
6. Must have a method for its analysis

Before approval, a compound classified as a food additive must have its safety established in experimental animal and/or human feeding trials. The regulations for each additive must describe the approved applications, amounts that are safe, and the conditions necessary to not harm the public. Approved additives can be found in the Code of Federal Regulations (CFR), Title 21, Parts 180–189. Some food additives fall into a category called generally recognized as safe or GRAS substances. The GRAS substances are exempt from food additive approval guidelines but still must be used only in approved products, within approved levels, and according to GMPs. All food additives, GRAS or not, must be listed on the food label. An effective active package that requires migration or has incidental migration would therefore need to have approval of the migrating compound as a food additive, and the label must declare that compound as a preservative.

OXYGEN SCAVENGERS/ANTIOXIDANTS

As stated in the introduction, the first patent for an oxygen scavenger for food was granted in 1938 for the removal of residual oxygen from the headspace of cans. The development of oxygen scavengers has continued with such advances as triggering the reaction by the presence of water, placing the scavenger in a film, and the development of non–iron based systems. Rooney (1995b) reported that 60 worldwide patents had been granted for oxygen scavenging sachets and 50 for oxygen scavenger–based polymers. The potential applications for oxygen scavenger plastics were summarized by Rooney (1995b) with the beverage applications including aseptically packaged liquids, bag-in-box beverages, coffee, and pasteurized drinks. For beverages, the use of oxygen scavengers in the sachet is not normally practical, thus closures (crowns) and polymers have had wider use. One problem facing packaging-based oxygen scavengers is stability with exposure to air prior to use. For blow-molded beverage containers, this can be overcome by combining the catalysts during the final blow-molding step closely followed by filling and sealing. The activating catalyst can also be combined with the substrate during filling, as is done with the Ox-Bar system. Other activating steps have also been developed such as exposure to water or light.

Oxygen scavenging had early application in the preservation of beer. Flavor quality was linked to oxygen content (Gray et al., 1948), leading the American Society of Brewing Chemists to recommend the study of adding antioxidants such as sulfites and ascorbic acid to retard flavor loss. Reinke et al. (1963) found that the use of cans lined with antioxidants improved beer shelf life. The removal of oxygen from the bottle headspace after sealing requires that a scavenger react with the gas without reacting with the beverage. To accomplish this, scavengers are incorporated into the closure (crown) by two methods. The first method utilizes a sachet attached to the inside of the closure with a membrane to separate the scavenger from the beer. The membrane permits oxygen and water vapor to permeate the sachet but prevents the scavenger from leaching into the beverage. The second method has a scavenger incorporated into a polymer coating on the inside of the closure. W.R. Grace developed a polymer liner for beer bottle caps containing sodium sulfate and sodium ascorbate in 1989. Polyvinyl chloride is often used as the carrier for the scavenger due to its high permeability to oxygen and water vapor. An oxygen-scavenging closure has been evaluated for use with several beer brands. The reaction rate of the ascorbate or erythorbate (ascorbate isomer) salts can be increased by the addition of transition metal salts. Copper and iron are the metals of choice, and this principle was applied by Zapat A (formerly Aquanautics Corporation) to produce Smartcap® in 1991. Smartcap and the newer version, Pureseal®, are produced by Zapat A, which sold over 1 billion crowns in 1993. The crowns were found to reduce oxygen levels in beer bottles after 1 to 3 months of storage with the effects maintained through 9 to 12 months of storage (Teumac, 1995). As of 1993, 20 microbreweries were believed to be using Pureseal crown liners including Sierra Nevada Brewing Co., Cellis Brewing

Co., Abita Brewing Co., and Full Sail Brewing Co. (Sacharow, 1995). The use of package oxygen scavengers for beer is gaining acceptance, allowing for maintenance of quality during shipment to more distant locations from the point of origin.

The use of scavengers for other beverages is being explored and is especially relevant for beverages containing natural colors and flavors that are susceptible to oxidation. Natural juices are susceptible to oxidation resulting in the loss of color, texture, flavor, and nutrients. Many beverages have been introduced that contain natural components or that have added nutrients that are oxygen labile. Some vitamins are very sensitive to oxidation, and the use of oxygen scavengers for beverages making health claims and containing oxygen-sensitive components may maintain nutritional quality.

The use of oxygen-scavenging sachets for beverages has been limited; however, oxygen-scavenging sachets have been used with roasted coffee. The Ageless E sachet (manufactured by Mitsubishi Gas Chemical Co.) contains ascorbic acid and absorbs oxygen and carbon dioxide. While oxygen is the main factor causing the deterioration of ground coffee, freshly ground coffee also releases significant amounts of carbon dioxide. To allow packaging of ground coffee almost immediately after grinding, sachets that absorb carbon dioxide are often added. Soft packs or pillow packs of ground coffee have been equipped with a one-way valve in the side of the package that opens and releases carbon dioxide when the internal pressure reaches a preset limit. This system facilitates the packaging of freshly ground coffee, minimizing exposure to oxygen while allowing for the release of carbon dioxide.

The addition of antioxidants to packaging has been shown to be effective in maintaining the quality of foods other than beverages. To prevent the oxidation of meat pigments, butylated hydroxytoluene (BHT) and butylated hydroxyanisole (BHA) were incorporated into polyethylene at the 0.1% level; BHT was effective in color maintenance (Dawson, 2001; Finkle et al., 2000). Both BHT and BHA migrated equally into ethanol (the standard Food and Drug Administration [FDA] fatty food simulant), while only BHT migrated into water. Table 10.1 shows the results of this experiment.

TABLE 10.1
Migration of BHA and BHT into Water and 95% Ethanol (ppm, w/v)

Antioxidant	Day 0	Day 3	Day 6	Day 9
BHA, water	0.83	4.03	9.62	18.45
BHT, water	0.00	0.00	0.00	0.00
BHA, 95% ethanol	1.22	19.51	26.13	25.32
BHT, 95% ethanol	0.00	0.00	0.00	0.00

This may have applications for beverages with labile components, and the use of natural antioxidants may need further investigation. Han et al. (1987) studied the diffusion of BHT from high-density polyethylene (HDPE) into packaged oat flakes and found that only 55% of the original BHT remained in the package after one week. Goyo Shiko (1993) patented the use of amino acids and saccharides in film coatings for their antioxidative properties. When heated, the proteins and simple sugars form brown pigments and antioxidants via the Maillard reaction. The film coatings were intended for beverage cans to be retorted with the retorting step used to catalyze the Maillard reaction and the antioxidant response.

ANTIMICROBIAL POLYMERS

Antimicrobial films can be divided into two general categories — those in which the antimicrobial agent migrates from the film and those in which the agent remains within the film material. Due to the nature of food, if the antimicrobial does not migrate from the film at least to the food surface, it will have limited effect. Several polymer materials have been developed that contain nonmigrating bactericides. These compounds are not yet approved as food additives and are not likely to be approved as such since the objective is to kill bacteria and other microorganisms coming in contact with the surface. This group of polymers is not designed to migrate from the surface into the environment or other contacting surfaces. One such compound is triclosan (5-chloro-2–2,4-dichlorophenoxy phenol), a chlorinated phenoxy compound. Triclosan has been used for 25 years as an ingredient in hospital soaps and dermatologic products. This compound inhibits the growth of a broad range of bacteria, molds, and fungi. The Microban Products Company has developed a process to incorporate triclosan into the structure of plastic polymers, opening the door to specialty applications that include surgical drapes, orthopedic cast liners, mattress/pillow covers, cutting boards, toothbrushes, children's toys, infant highchairs, shower curtains, toilet/door handles, mops, mop handles, and paint. Triclosan has also been used as an ingredient in toothpaste. Triclosan is incorporated into the molecular spaces that exist in a plastic polymer and is available in polypropylene, polyethylene, polybutyl terephthalate, and other polymeric materials.

Another antimicrobial compound that has been incorporated into surfaces is silver. Surfacine Inc. reports that silver is a safe biocide with no human toxicity. Silver has been incorporated into zeolite (a hydrated aluminosilicate with an open three-dimensional crystal structure in which water is held in the cavities of the lattice). The water can be driven off by heat, and the zeolite can absorb other molecules. The silver-treated zeolite has been incorporated into a polymer film and will be discussed in more detail later in the chapter.

Benzoic anhydride has been incorporated into low-density polyethylene films to inhibit mold growth. Quaternary ammonium salts (quats) have also been added to acrylic resins. These are proposed for use in prostheses, dental bridges, and adhesives. Most of these products are not approved in the U.S. as food additives; thus, most are not currently used in food packaging. They may have some application for processing surfaces where cross-contamination is a problem.

The second category of film with migrating antimicrobials must be concerned with the effect on the food of the migrating species. Some bacteriocins and enzymes are approved as food additives and thus may be effective for use in migrating antimicrobial films. Nisin is a bacteriocin approved for use in cheese spread and liquid egg in the U.S., with wider approval in other countries. Glucose oxidase is an enzyme that produces hydrogen peroxide, which destroys bacterial cells upon contact. Lysozyme is found naturally in milk and egg white and in a slightly different form in human tears. Lysozyme destroys cell membranes of bacteria but, like nisin, it is limited in effectiveness to Gram-positive bacteria since Gram-negative bacteria have an additional outer cell membrane that blocks access to the enzymes' and bacteriocins' active site. The Japanese report the development of IR-emitting films by the incorporation of radiation emitters into film materials. This option is the least developed and documented at this point. A short list of antimicrobials available for use in films is shown in Table 10.2.

Two approaches can be taken to produce an antimicrobial film. A film surface can be coated with an antimicrobial, or the antimicrobial can be incorporated into the film material. Each approach has its advantages and disadvantages. Coating a package surface allows quick release of the antimicrobial, and the antimicrobial itself does not interfere with the film structure. This can be a concern especially in synthetic polymer films, which are

**TABLE 10.2
A Short List of Antimicrobials Available for Use in Polymer Films**

Antimicrobial Category	Examples
Organic acids	Salt, acid, anhydride
Natural derivatives	Spice extracts
Enzymes	Lysozyme, glucose oxidase
Bacteriocins	Nisin, pediocin
Chelators	EDTA, citric acid
Gases	CO_2, ozone, chlorine oxide
Silver	Ions, salts

often nonpolar, since many of antimicrobials are polar. Incorporation of the antimicrobial into the film material must take into consideration the effect on the package properties, but a continued release of the antimicrobial into the food at the film surface can be achieved. Often, the determining factor in which approach to take lies in the objective of the application. A rapid and immediate release of a coating into the food bulk might be achieved more economically by the direct addition of the antimicrobial to the food. The cost of coating a film when the effect is likely to only last several minutes to hours might not be the best option. A reduction in initial bacteria, mold, or fungi numbers could and probably should be addressed prior to packaging. The incorporation of the antimicrobial can give extended suppression of microbial growth well into the distribution and handling cycle for processed foods having a longer shelf life. The focus of this discussion will be on films with the antimicrobial incorporated into the film structure.

Research has been conducted on both biopolymer and synthetic polymer films with antimicrobials incorporated into their structure. Films containing silver appear to have the most interest at present. Some metals such as silver and copper are toxic to microorganisms and viruses when the metal in ion form comes in contact with them. Copper is not concentrated in higher animals, which makes it safe compared to some metals, but nevertheless copper is regarded as toxic and is not permitted to be used in contact with food. Copper is also a prooxidant and thus can accelerate the deterioration of food quality. Silver ions have the strongest antimicrobial activity among metals (Brody, 2001b) but the ion is not released as easily as that of copper. Thus, silver's antimicrobial activity is not as strong as that of copper in the nonionic or salt state. Silver is used in water treatment, and the silver nitrate form is used as an antiseptic in hospitals. Silver is believed to interfere with the electron transport functions of microorganisms and with mass transfer across cell membranes. Silver has a broad spectrum of activity against both aerobic and anaerobic bacteria; however, some resistant strains that absorb silver have been found.

Antimicrobial packaging using silver has employed zeolite as the carrier. The zeolite retains the silver ions in a stable and active form to make the metal more effective. Once released, silver ions will react with organic metal compounds such as sulfur to make them inactive. Thus, the silver is most effective when retained in the zeolite structure, and the bacteria must come in contact with the package surface for the most potent killing effect to occur. Due to expense, silver–zeolite is incorporated into plastics as a thin (3–6 μm) laminate layer at the food contact surface. The normal incorporation level is 1–3% (Brody, 2001b). Three amino acid types affect the diffusion of silver from zeolite. Glycine-type (polar–uncharged), lysine-type (positively charged) and cysteine type (sulfur-containing) amino acids all increase the release of silver ions from zeolite. Lysine and cysteine form strong associations with silver, thus inhibiting its antimicrobial activity once released from zeolite. Glycine forms a weak association that does not prevent silver from acting on microorganisms; this may increase the activity of the ion by stimulating its release from the carrier. DuPont markets a powder, MicroFree®, designed to impart antimicrobial properties to film when added to the resin. Three powders are offered; all are inorganic, nonvolatile, and stable to light and heat. MicroFree uses silver ions (bactericide), copper oxide (fungicide), and zinc silicate (fungicide), with various support vehicles for different applications. The types are Z-200 (silver on a zinc oxide core), T-558 (silver, copper oxide, and zinc silicate on a titanium dioxide core), and B-558 (silver, copper oxide, and zinc silicate on a barium sulfate core). Another silver–zeolite antimicrobial powder designed to be added to resin is Zeomic from Shinanen New Ceramics Co. Many antimicrobial package types are available in Japan. Examples are Apacider-A® from Sangi, which uses silver bonded to calcium phosphate on zeolite, and a low-density polyethylene film with zeolite produced by Tadashi Ogawa. The film is touted to trap microorganisms in the zeolite pores and trap ethylene gas to preserve respiring plant tissue. Ogawa also claims that the film absorbs IR and reemits it at a frequency that is bactericidal. Silvi film from Nimiko Co. uses a silver ion and silica–oxide blend in plastic film to inhibit bacterial and mold growth. The gradual release of silver oxide from the film is reported to be effective in fresh meat, respiring vegetable, and liquid food systems.

A long-term preservative pouch for drinking water called Miracle Water Pack® was developed jointly by the Try and Taiyo chemical companies. The pouch has five nylon/polyethylene layers with the inner food contact layer impregnated with silver zeolite. Traditional zeolite contains pores that are large enough to impart a cloudy appearance to a clear film. The unique feature of Miracle Water Pack is the transparency of the film, attributable to the use of zeolite with smaller-diameter pores. Bottled water requires a transparent container to allow for visual inspection of the product. Benomyl (a fungicide) is another additive in resin-based food packaging material available in Japan that inhibits mold growth on food. Sorbic acid has also been used as a coating and as part of wraps or films to inhibit mold growth on foods.

Natural antimicrobials that have been utilized in packaging applications include spice extracts, bacteriocins, chlorine dioxide gas, ethanol, and wasabi (a derivative from Japanese horseradish). Only a handful of commercial films using "natural" antimicrobials have been discussed in the literature (Table 10.3); however, numerous research papers report testing antimicrobial packaging using natural products.

The bacteriocin nisin is one of the more researched and effective antimicrobials. Nisin is a polypeptide that lyses bacterial cells by interacting

TABLE 10.3
Packaging Materials Using Natural Antimicrobials

Sponsor	Antimicrobial	Application
Viskase	Bacteriocins	Meat casings
Bernard Technologies	Chlorine dioxide	Meat
Freund Ind. Co. Ltd.	Ethanol	Bakery items
Sekisui Jushi	Wasabi (allylisothiocyanate)	Lunch boxes, wraps

with sulfur-containing cell membrane compounds. Nisin is normally ineffective against Gram-negative bacteria, since they possess an outer cell membrane that blocks the active site. This can be overcome by the combination of nisin with food-grade chelators such as EDTA and citric acid. Polyethylene films and corn zein films were shown to reduce *Listeria monocytogenes* populations in peptone water from 8 logs ([colony forming units] cfu/ml) to below detectable levels ($<10^2$) after 24 hours (Hoffman et al., 1997, 2001). Corn zein films impregnated with nisin reduced *L. monocytogenes* in skim milk by 3 logs (cfu/ml) after 48 hours. The diffusivity of nisin-impregnated corn zein and wheat gluten films into water were determined for both cast and heat-pressed films (Teerakarn et al., 2001). The cast wheat gluten film had the greatest diffusivity, while the cast corn zein film had the lowest (Figure 10.1). The heat-pressed wheat gluten and corn zein films did

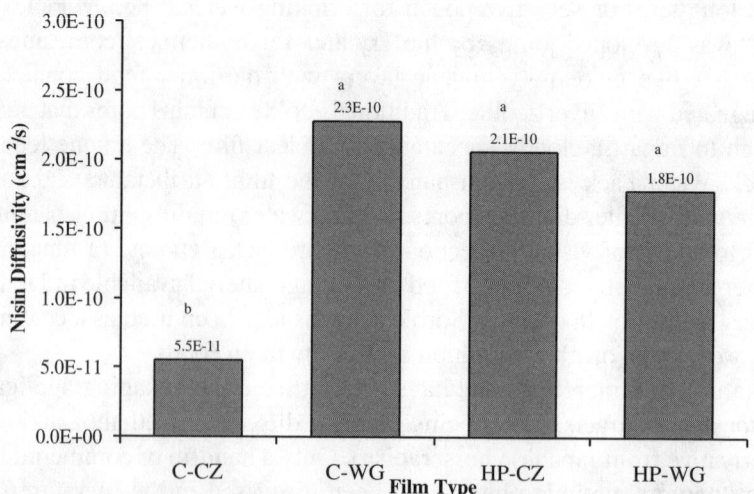

FIGURE 10.1 Diffusivity of cast corn zein (C-CZ), cast wheat gluten (C-WG), heat-pressed corn zein (H-CZ), and heat-pressed wheat gluten (H-WG) films exposed to water at 5, 25, 35, and 45°C. [a,b]Values with the same superscripts were not significantly different ($p > 0.05$).

not differ in diffusivity. The nisin diffusivity of each of these four film types adhered to Fickian diffusion giving a straight line fit for Arrhenius plots between 5 and 40°C. Nisin is approved in the U.S. for direct addition to liquid egg and processed cheese and has wider approval for use in foods in other countries. Therefore, the use of nisin and other components in packages for extended shelf life beverages may have promise.

BIO-BASED MATERIALS FOR PACKAGING

One of the leading research units for bio-based food packaging materials is The Royal Veterinary and Agricultural University in Denmark. Researchers there are developing starch-based materials suitable for packaging beverages as well as other food products. Biopolymer beverage packages have been developed using polylactate (PLA) and polyhydroxy-alcanoates (PHA) (Haugaard and Bertelsen, 2001). Cargill Dow's NatureWorks PLA® and Mitsui's LACEA® are current packaging materials based on PLA. Hycail also supplies a PLA-based product. Biomer sells a product under the same name that is based on PHB. Novamont (Mater Bi®), Biotec (Biolast®), and Earth Shell (Earth Shell®) also sell starch-based polymer packaging materials. When fresh orange juice was stored for 7 days at 4°C in containers made from polyethylene (PE), PLA, and polystyrene (PS), the PLA maintained the yellow color better than PE and equally as well as PS. In addition, after 7 days, vitamin C content in juice stored in PLA dropped from 54 to 52 mg/100g compared to 54 to 50 mg/100g for PS and PE. The PLA and PS packages showed no detectable scalping of D-limonene, while PE scalped 15 mg/package. The conclusions drawn by Haugaard and Bertelsen (2001) were that PLA and PHB are suitable for packaging orange juice as well as other beverages. These recent advances also offer the opportunity to use bio-based materials in active packaging applications.

TAINT REMOVERS

Flavor scalping by plastics is a well-documented phenomenon. One example of flavor scalping in beverages is limonene scalping from orange juice by surlyn and polyethylene. In aseptic packages stored at 24°C for two weeks, 30% of the limonene from orange juice was found in the surlyn layer and 20% in the polyethylene layer (Hirose et al., 1989). Although plastics have not been used to selectively remove specific off-flavors in beverages, this principle has been tested in orange juice to remove a bitter compound, limonin. Limonin is concentrated in juice during the extraction and pasteurization of fresh juice. Chandler and Johnson (1979) showed that a 1-liter plastic bottle coated with cellulose acetate or acetylated paper reduced the limonin content in juice from 42 to 11 mg/kg after three days.

Other methods reported to remove off-flavors address amines from protein degradation and aldehydes from lipid oxidation. Amines are strong bases and react readily with acids. Food acids such as citric acid were added to heat-extruded polymers to absorb the amines (Hoshino and Osanai, 1986). The ANICO BAG, produced by ANICO Company Limited Japan, contains an iron salt with organic acids in films to absorb and oxidize amine compounds. DuPont Polymer Packaging Division produces a high-density polyethylene resin (Bynel IX101) in an intermediate layer of film that is claimed to remove hexanal and heptanal from foods (Dupont Polymers, 1993). Taint removers are an active packaging type that is likely to see further development in the future. M.L. Rooney (1995b) states, "A fertile research field would seem to be open especially with liquid food since solubility and diffusion of food constituents in the packaging can be utilized so that the removal process is not limited to compounds with a significant vapor pressure at distribution temperature." He further states that the taint removers must not conceal low-quality or unsafe foods.

CONCLUSION

Active packaging for beverages is currently used in the form of oxygen scavengers, particularly for the bottle crowns in specialty beers. The more widespread use of this and the other active packaging types discussed in this chapter with beverages seems likely in the future. As more beverages are marketed with natural, fresh, and health-related claims, active packaging is likely to play a role in maintaining the quality of these products.

REFERENCES

Brody, A.L., What's the hottest food packaging technology today? *Food Technol.*, 55, 82–84, 2001a.

Brody, A.L., Antimicrobial packaging, in *Active Packaging for Food Applications*, Brody, A.L., Strupinsky, E.R., and Kline, L.R., Eds. Technomic Publishing Company, Lancaster, PA, 2001b, pp. 131–189.

Chandler, B.V. and Johnson, R.L., New sorbent gel forms of cellulose esters for debittering citrus juices, *J. Sci. Food Agric.*, 30, 825–832, 1979.

Dawson, P.L., Active Packaging: Films and Coatings for Extended Shelf Life, International Animal Agriculture and Food Science Conference Proceedings, Abstract No. 426, 2001, p. 103.

DuPont Polymers, Bynel IPX101, interactive packaging resin, in *Active Food Packaging*, Rooney, M.L., Ed., Blackie Academic and Professional, New York, 1993, p. 101.

Finkle, M.E., Han, I.Y., and Dawson, P.L., Effect of Antioxidant Impregnated Films on the Color of Beef, International Congress on Meat Science and Technology, Buenos Aires, August 2000.

Goyo Shiko, K.K., Packaging Material with Good Gas Barrier Property and Oxygen Absorbing Property Comprises High Molecular Substance Amino Acids and Hydroxyl Groups Containing Reducing Resin, Japanese Patent 5186635, 1993.

Gray, P., Stone, I., and Atkin, L., Systematic study of the influence of oxidation on beer flavor, *Am. Soc. Beer Chem. Proc.,* 101–112, 1948.

Han, J.K., Miltz, J., Harte, B.R., Giacin, J.R., and Gray, J.J., Loss of 2-tertiary-butyl-4-methoxy phenol (BHA) from high-density polyethylene film, *Polymer Eng. Sci.,* 27, 934–938, 1987.

Harima, Y., *Food Packaging,* Academic Press, London, 1990, pp. 229–252.

Haugaard, V.K. and Bertelsen, G., Potential for Bio-based Materials for Food Packaging, International Animal Agriculture and Food Science Conference Proceedings, Abstract No. 430, 2001, pp. 103.

Hirose, K., Harte, B.R., Giacin, J.R. Miltz, J., and Stine, C., Sorption of α-limonene by sealed films and effect of mechanical properties, in *Food and Beverage Packaging Interactions,* Hotchkiss, J.H., Ed., ACS Symposium Series No. 365, American Chemical Society, Washington, D.C., 1989, pp. 28–41.

Hoffman, K.L., Dawson, P.L., Acton, J.C., Han, I.Y., and Ogale, A.A., Film formation effects on nisin activity in corn zein and polyethylene films, *Res. Dev. Activ. Rep. Military Food Packaging Syst.,* 47, 203–210, 1997.

Hoffman, K.L., Han, I.Y., and Dawson, P.L., Antimicrobial effects of corn zein films impregnated with nisin, lauric acid, and EDTA, *J. Food Prot.,* 64, 885–889, 2001.

Hoshino, A. and Osanai, T., Packaging Films for Deodorization, Japanese Patent 86209612, 1986.

Reinke, H., Hoag, L., and Kincaid, C., Effect of antioxidants and oxygen scavengers on the shelf-life of canned beer, *Am. Soc. Beer Chem. Proc.,* 175–180, 1963.

Rooney, M.L., Overview of active food packaging, in *Active Food Packaging,* Rooney, M.L., Ed., Blackie Academic and Professional, New York, 1995a, p. 1.

Rooney, M.L., Active packaging in polymer films, in *Active Food Packaging,* Rooney, M.L., Ed., Blackie Academic and Professional, New York, 1995b, pp. 94–107.

Sacharow, S., Commercial applications in North America, in *Active Food Packaging,* Rooney, M.L., Ed., Blackie Academic and Professional, New York, 1995, pp. 203–214.

Teerakarn, A., Hirt, D.E., Rieck, J.R., Acton, J.C., and Dawson, P.L., Nisin diffusion in protein films: effects of time and temperature, *J. Agric. Food Chem.,* 2003 (submitted for publication).

Teumac, F.N., The history of oxygen scavenger bottle closures, in *Active Food Packaging,* Rooney, M.L., Ed., Blackie Academic and Professional, New York, 1995, pp. 193–201.

Wagner, J., The Advent of Smart Packaging, *Food Eng. Int.,* Dec. 1989, p. 11.

SUGGESTED READING FOR MORE INFORMATION

Brody, A.L., Strupinsky, E.R., and Kline, L.R., Eds., *Active Packaging for Food Applications,* Technomic Publishing Co., Inc., Lancaster, PA, 2001.

Robertson, G.L., Ed., *Food Packaging: Principles and Practice,* Marcel Dekker, Inc., New York, 1993.

Rooney, M.L., Ed., *Active Food Packaging,* Blackie Academic and Professional, New York, 1995.

Index

A

ACE, *see* Angiotensin-I converting enzyme
Acetobacter, 98, 99
Acid sanitizers, 185
Adenosine triphosphate (ATP), 189, 190
Aflatoxin production, caffeine and, 105
Age-related macular degeneration (ARMD), 25, 36, 51
Air filters, 191
Alcohol, 24
Alicyclobacillus, 98
  acidoterrestris, 99, 100
  spoilage, occurrence of, 100
Allergens, removal of, 13
Aloe, 47
Alternative processing technologies, 81
Alzheimer's disease, 48
Amino acids, branched-chain, 26
Angiotensin-I converting enzyme (ACE), 59
Anthocyanidins, 52
Anthraquinones, 32
Antiinflammatory agents, 54
Antimicrobial(s)
  natural, 213, 214
  packaging, 212
  polymers, 210
Antinutrients, removal of, 13
Antioxidant(s), 42, 53
  natural, 14
  nutraceuticals acting as, 54
Anxiety, 49
Apple cider
  Cryptosporidium and, 108, 126
  manufacturers, conditions observed in FDA inspections of fresh unpasteurized, 114
  time–temperature conditions for pasteurization of, 80
Apple juice, unpasteurized, 113
Apples, dropped, 138
ARMD, *see* Age-related macular degeneration
Aseptic packaging, 138
Asian ginseng, 29, 46
Aspartame, 29, 45
Athletic events, fluid intake and, 39
ATP, *see* Adenosine triphosphate

Aureobasidium, 101
Aurones, 33

B

Bacillus, 98
  cereus, 104
  coagulans, 99
  licheniformis, 99
  macerans, 99
  polymyxa, 99
  subtilis, 22, 99
Bacteria
  ability of to resist stress, 79
  E. coli, 55
  Gram-negative, 211
  indicator, 102
  intestinal, 43
  lactic acid, 20, 99
  nonthermal processing methods of inhibiting pathogenic, 83–84
Bakery products, active packaging of, 205
Beer, 24
  phenolics in, 25
  yeast, dietary fiber from, 20
Benign prostatic hyperplasia (BPH), 46
Benzoic anhydride, 211
Benzoquinones, 32
Beverage(s)
  accolades of, 23
  carbohydrate–electrolyte, 38
  case studies for, 190
  consumption, U.S. per capita, 97
  containers, blow-molded, 208
  new, 96
  nonbovine milk-related, 27
  organic, 197
  packages, biopolymer, 215
  products, consumer acceptance of, 16
  sport, 38, 39
  stimulator, 40
  wellness, 41
BHA, *see* Butylated hydroxyanisole
BHT, *see* Butylated hydroxytoluene
Bifidobacterium
  acidophilus, 21

bleve, 20
lactis, 20
longum, 20
Biflavonoids, 33
Bil Mar Foods, 177
Biodiversity, 193
Bioflavonoids, 52
Biopolymer beverage packages, 215
Biotransformation enzymes, 58
Birth defects, neural tube, 17
Black cohosh, 29, 47
Black tea, CHD and, 28
Blow-molded beverage containers, 208
Body Mass Index, 45
Botanicals, 46
Botrytis aclada, 88
Bottled water, 182
BPH, *see* Benign prostatic hyperplasia
Brassica, 53, 58
Breast cancer, 25, 52
Brettanomyces intermedius, 101
Broccoli
    glycoalkaloids in, 11
    sulfurofane in, 14
Butylated hydroxyanisole (BHA), 209
Butylated hydroxytoluene (BHT), 209
Byssochlamys, 101
    fulva, 102
    nivea, 102

C

Caffeine, 39, 105
Calcium
    carbonate–citric acid–malic acid (CCM), 50
    osteoporosis and, 16, 50
Campylobacter, 127
Cancer, 98
Candida
    famata, 103
    guillermondii, 103
    krusei, 103
    parapsilosis, 103
Carbohydrate(s), 29
    –electrolyte beverages, 38
    nondigestible, 18
    subcategories of, 46
Carnitine, 26
β-Carotene, 36, 51
Carotenoids, 31, 36, 53, 54
Carrageenan, 30, 45
Casein dodecanoic peptide, 21
Cataracts, 25, 36, 51
Catechins, 42

CCM, *see* Calcium carbonate–citric acid–malic acid
CCP, *see* Critical control point
CDC, *see* Centers for Disease Control and Prevention
Cell kinetics, 42, 56
Centers for Disease Control and Prevention (CDC), 106
Centrum voor Kwaliteitscontrole (CKC), 3
Cerebrosides, 31
Certification, pros and cons of, 6
CFR, *see* Code of Federal Regulations
CGMPs, *see* Current good manufacturing practices
CHD, *see* Coronary heart disease
Chitosan, 21
Chocolate, 28
Cholesterol, 24, 46
Chromium picolinate, 51
Chromones, 32
Cider
    /juice production, best practices for, 115
    unpasteurized, 81
Citrus fruits, 35
Citrus Hill, 50
Citrus juices, analysis for *E. coli* in, 152
CKC, *see* Centrum voor Kwaliteitscontrole
Cladosporium, 101
Cleaning
    compounds, 180
    efficacy of, 189
    most important key to, 185
Clean-out-of-place (COP) systems, 185
Clostridium, 98
    botulinum, 128
    butyricum, 99
    pasteurianum, 99
CLs, *see* Critical limits
Coca-Cola Classic®, 40
Cocoa, 28
    nutraceutical compounds found in, 28
    phenolics in, 35
Code of Federal Regulations (CFR), 207
Codex Alimentarius Commission, 143
Codex Committee on Food Hygiene, 1, 4, 181
Coffee, 28
    nutraceutical compounds found in, 28
    phenolics in, 35
Coloring agents, 38
Colostrums, 27
COP systems, *see* Clean-out-of-place systems
Corn, hybrid seed, 10
Coronary heart disease (CHD), 25
    black tea and, 28
    green tea and, 28

Index

mortality, 25
prevention of, 36
wine consumption and, 25
Coumarins, 32
Cow's milk, lactoferrin in, 27
Critical control point (CCP), 142, 163, 167
   corrective actions for, 169
   decision tree, 167
   monitoring of, 170
Critical limits (CLs), 143, 167
Crop(s)
   genetically modified, 9, 10
   herbicide-tolerant, 10
   injury, reduced, 10
Cross-contamination, 172, 198
Cryptosporidium, 181
   apple cider and, 108
   parvum, 86, 103, 107
Current good manufacturing practices (CGMPs), 132, 136, 141, 145
Cyclospora, 109

D

DADS, *see* Diallyl disulfide
Daidzein, 25
DDT, banning of, 194
Decision tree, critical control point, 167
Defect action levels, 182
Degree of polymerization (DP), 50
Dekkera
   bruxellensis, 101
   nardenensis, 101
Depression, 46, 49
Detergents, 185
Detoxifying enzymes, 58
Dextrin, indigestible, 20, 22
Diabetes, 98
Diallyl disulfide (DADS), 53
Dietary fiber, 20, 50
   bulking sources of, 57
   history of, 46
   isolated sources of, 30
Dietary Reference Intakes (DRIs), 17, 46
Dietary Supplement Health and Education Act (DSHEA), 18
Dihydrochalcones, 33
Dihydroflavonol, 33
Disease prevention, 16
DNA
   fingerprint pattern, 126
   GM, 13
   susceptibility of to damage, 56
DP, *see* Degree of polymerization

Drinking water regulations, 182
Drinks, 38
DRIs, *see* Dietary Reference Intakes
Dropped apples, 138
DSHEA, *see* Dietary Supplement Health and Education Act
DuPont, 213, 216

E

Echinacea, 46, 47
EGEC, *see* Enterohemorrhagic *E. coli*
Eicosanoids, 57
Electron beam sterilization, 139
Elements, essential, 50
ELISAs, *see* Enzyme-linked immunosorbent assays
Embryo rescue, 10
Employee hygiene, 116
Enteral formulas, 26
Enterobacter aerogenes, 77
Enterohemorrhagic *E. coli* (EHEC), 87, 106
Environmental Protection Agency (EPA), 11, 131, 184, 195
Enzyme(s)
   activity, 58
   angiotensin-I converting, 59
   biotransformation, 58
   detoxifying, 58
   -linked immunosorbent assays (ELISAs), 12
   P450 mixed-function oxidase, 59
EPA, *see* Environmental Protection Agency
Ephedra, 47
Epicatechin, 25
Equipment sanitation, 116
Escherichia coli
   analysis for in citrus juices, 152
   bacteria, urinary tract–pathogenic, 55
   enterohemorrhagic, 87, 106
   inactivation of on surface of oranges, 78
   O157:H7, 86, 98, 107, 118
      effects of chemicals in reducing, 75
      infections, 126
      outbreaks of, 110
      waterborne transmission of,
   potassium sorbate and, 78
Essential elements, 50
Ethanol-releasing sachets, 206
Eupenicillium, 101
EUREGAP certification protocols, 2, 6
Eusiderin, 33
Evening primrose, 47
Exercise, 26

F

Fagopyritols, 29
Fat substitutes, 29, 46
FDA, *see* Food and Drug Administration
Federal Food Drug, and Cosmetic Act, 144, 147, 131
Fermentable sugars, 73
Fermentation, 110
Fermented foods, 27
Feverfew, 47
Fibre Mini, 18
Fitness Water, 27
Flavonoids, 32
Fluid Meal Replacements, 25, 26
Folic acid, acceptance of, 17
Food(s)
  acids, 216
  additives, 45
  for Specified Health Use (FOSHU), 18, 19
  hazard, definition of, 144
  labeling, 207
  medical, 26
  quality, improvement of attributes of, 17
  recommendations for consumption or avoidance of specific, 16
  safety and quality pyramid, 176
  safety hazards, illness from, 109
  supply, uncovered, 1
Foodborne illness
  juices as vehicle for outbreaks of, 128
  microorganisms related to, 104
Food and Drug Administration (FDA), 2, 9, 11, 18, 74, 98, 125, 177
  fatty food simulant, 209
  final rule on use of HACCP system, 158
  Generally Recognized as Safe status, 44
  GMP matrix example, 178–179
  guideline for minimizing microbiological hazards, 79
  inspections, 114, 135, 136
  public meeting, 132
  recall data, 128
  sampling survey, 130
  surveillance strategies, 136
Fortified Water, 27
FOS, *see* Fructooligosaccharide
FOSHU, *see* Foods for Specified Health Use
Free radical formation, prevention of, 54
French paradox, 25
Fructooligosaccharide (FOS), 20, 22, 45, 50, 58
Fruit(s)
  contamination of, 74
  drinks, recalled, 129
  effects of chemicals in reducing bacteria on surface of, 75–77
  failure to wash, 78
  juice, spoilage in, 99
Fumonisin-producing fungi, 10
Functional foods, *see* Nutraceuticals, beverages as delivery systems for
Fungi, fumonisin-producing, 10
Fusarium
  graminearum, 57
  venenatumon, 57

G

Galactooligosaccharides (GOS), 27, 55
GALT, *see* Gut-associated lymphoid tissue
Gangliosides, 31
GAPs, *see* Good Agricultural Practices
Gastric ulcers, 55
Gatorade®, 39
Gene expression, 55
Generally Recognized as Safe (GRAS), 44, 207
Genetically modified (GM)
  crops, 9, 10
  DNA, 13
  foods, output traits, 13
  sampling, confidence in adequacy of current, 13
Genetically modified organisms (GMOs), 9–14
  future of genetically modified foods, 13–14
  history of genetic modification of food plants and animals, 10
  identity preservation and international market, 11–13
    detection of genetically modified ingredients, 12
    difficulties with product labeling, 12–13
    regulation of genetically modified crops, 10–11
Genistein, 25
Giardia, 181
Ginger, 47
Ginkgo biloba, 29, 46, 48
Ginseng, 48
Globin hydrolysate, 23
Gluconobacter, 99
Glucose oxidase, 211
Glucosinolates, 58
Glutamine, 26
Glutathione, 34, 59
Glycoalkaloids, 11
GM, *see* Genetically modified
GMOs, *see* Genetically modified organisms
GMPs, *see* Good Manufacturing Practices

Index

Goldenseal, 29
Good Agricultural Practices (GAPs), 2, 7, 79
   adoption of, 6
   Third World perception of, 3
Good Manufacturing Practices (GMPs), 159, 177
   General Provisions within, 179
   pest control measures following, 199
GOS, *see* Galactooligosaccharides
Gram-negative bacteria, 211
GRAS, *see* Generally Recognized as Safe
Green tea, CHD and, 28
Guava leaves polyphenol, 22
Gum arabic, 30
Gut-associated lymphoid tissue (GALT), 54

H

HACCP (Hazard Analysis and Critical Control Point), 1, 126
   control measures, 140
   failure, 137
   mandatory, 118, 134, 141
   model, 116, 117
   plan, 146
   prerequisite programs, 2
   records, 149
   summary table, 167
   system
      purpose of, 98
      validation of, 143
   training, 150
   verification, 148, 170
HACCP, applied approach, 157–173
   benefits of HACCP system, 158–159
   definitions, 161
   development of HACCP plan, 162–163
      assembling of HACCP team, 162
      description of food and distribution, 162
      description of intended use and consumers of food, 162
      development of flow diagram describing process, 162–163
      flow diagram verification, 163
   HACCP to control hazards, 158
   hazard components, 159
   principles of HACCP, 163
      determination of critical control points, 167
      establishment of corrective actions, 169–170
      establishment of critical limits, 167–168
      establishment of monitoring procedures, 168–169
      establishment of verification procedures, 170–171
      hazard analysis, 163–167
      records, 171–173
   supporting programs, 159–160
HACCP, final rule, 125–156
   concerns with juice, 126–135
      FDA public meeting, 132–135
      illnesses from hazards not heat treatable, 128–130
      microbial outbreaks, 126–128
      pesticides, 130–132
      underreporting, 130
   consideration of how to address juice concerns, 135–153
      alternatives, 136–141
      current inspection system, 135–136
      current regulation of juice, 135
      decision to mandate HACCP, 141–143
      final rule, 143–153
Handwashing, 180
Hawthorn, 29, 48
Hazard(s)
   analysis, 145, 164
   identification of potential, 164
Hazard Analysis and Critical Control Point, *see* HACCP
HDL, *see* High-density lipoproteins
Health benefits, new products claiming, 98
Heart health, 98
Heat
   pasteurization, 80
   -resistant bacilli, 100, 101
Helicobacter pylori, 55
Hemolytic uremic syndrome (HUS), 106, 126
Herbal medicines, 47
Herbicide tolerance, 9, 10
Hesperetin, 35
High-density lipoproteins (HDL), 34
High pressure processing (HPP), 82, 83
HIV, *see* Human immunodeficiency virus
Homemade chicken soup, 19
Hormonal action, 56
Horticultural products
   best practices for global production of, 2
   smart films used in, 205
Hot flashes, 47
HPP, *see* High pressure processing
Hudson Industries, 177
Human immunodeficiency virus (HIV), 48
HUS, *see* Hemolytic uremic syndrome
Hydrolyzed guar gum, 20
Hydroxyeicosatetraenoic acid, 57

I

Identity preservation (IP), 9, 11, 12
IFT, *see* Institute of Food Technologists
Illness, emerging pathogens and outbreaks of, 105
Immune enhancing nutrients, 57
Immune stimulators, 43, 57
Independent Organic Inspectors Association (IOIA), 202
Indigestible dextrin, 20, 22
Insect resistance, 9, 10
Insomnia, 49
Institute of Food Technologists (IFT), 26, 81
International Organization for Standardization (ISO), 2
Intestinal absorption, 43, 58
Intestinal bacteria, energy for, 43
Intestinal bulk, 57
IOIA, *see* Independent Organic Inspectors Association
IP, *see* Identity preservation
Iron absorption, 51
Irradiation, 84, 87
Irritable bowel syndrome, 49
ISO, *see* International Organization for Standardization
ISO 9000, 6
Isothiocyanates, 43, 53
Isotonic drinks, 38

J

Japanese horseradish, 213
Japanese paradox, 25
Juice(s)
  beverage, naming of nonstandardized, 23
  definition of, 35, 95
  non–heat-treatable hazards in, 129
  processes, nonthermal, 89
  processors, best practices for, 113
  products
    foodborne illness attributed to, 4
    food hazards associated with, 126
  regulations, 183
Juices and juice products, ensuring safety in, 1–7
  certification, 5–6
  evolution of GAPs, 2–3
  microbiological and chemical safety, 3–5
  proactive approach as good business, 6
Juice processing, organic alternative, 193–203
  certifiers and certification process, 201–203
  federal regulations, 200–201
  future of organic products, 203
  history of organic movement, 194
  market demand for organic products, 194–195
  organic processing and regulations, 195–197
  processing of organic product, 198–200
  sanitation, 197–198

K

Kava-kava, 24, 29, 48
Kefir, 27
Kool-Aid®, 38

L

Labeling requirement, 139
Lactic acid bacteria, 20, 99
Lactobacillus, 50, 99
  acidophilus, 20
  casei, 20, 21
  helveticus, 21
Lactoferrin, 27
Lacto-tri-peptide, 21
Lavinose, 20
Laxative, 47
LDL, *see* Low-density lipoproteins
Leuconostoc, 99
Leukotrienes, 57
Licorice root, 29, 48
Life Plan®, 41
Lignans, 33, 52
Lipid(s), 42
  complex, 30
  hormones, 57
  simple, 30
  structured, 31
Lipovitan®, 41
Lipoxins, 57
Listeria monocytogenes, 84, 86, 133, 111, 214
Living drugs, 52
Low-density lipoproteins (LDL), 25, 34
Lutein, 25, 36, 37, 42
Lycopene, 14, 30, 37, 42

M

Maltitol, 22, 29
Maltodextrin, 45
Mannitol, 29
Meat factor effect, 43, 51
Mechanical scrubbing, 74
Medical foods, 26
Medium-chain triglycerides, 29
Memorandum of understanding (MOU), 150
Memory, 46, 48

Metamucil®, 58
Microban Products Company, 210
Microbial contamination, control of, 74, 78
Microbiology of fruit juice and beverages, 95–123
   early outbreaks, 105–106
   emerging pathogens and outbreaks of illness, 105–110
      illness from other potential food safety hazards, 109–110
      outbreaks in 1990s, 106–109
   ensuring safety of juice, 110–118
      GMP and best practices for juice processors, 113–116
      model HACCP, 116–118
   microbial spoilage of fruit and fruit juice and beverages, 98–105
      bacteria, 98–101
      indicator bacteria and pathogenic organisms, 102–103
      mycotoxins, 105
      pathogenic yeasts, 103
      protozoa, 103
      viruses, 103–104
      yeasts and molds, 101–102
Microwaves, use of to heat food for commercial pasteurization, 87
Migraine headaches, 47
Milk
   -derived ingredients, 11
   lactoferrin in, 27
   soy, 27, 51
Milk thistle, 48
Model HACCP, 116, 117
Mold(s)
   contamination, 101, 110
   heat-resistant genera of, 101
Monilia fructigena, 88
Monitoring
   equipment, calibrated, 168
   programs, effectiveness of, 171
Monoterpenes, 31
Motion sickness, 47
MOU, *see* Memorandum of understanding
Mucilages, 30
Mycotoxins, 10, 105

N

NACMCF, *see* National Advisory Committee on Microbiological Criteria for Foods
Naftoquinones, 32
NAI, *see* No Action Indicated

Narigenin, 35
Natamycin, 129
National Academy of Sciences, UL defined by, 24
National Advisory Committee on Microbiological Criteria for Foods (NACMCF), 111, 133, 134, 138
National Organic Program, 196
National Organic Standards Board, 198
Natural antimicrobials, 213, 214
Natural toxicants, removal of, 13
Neosartorya, 101, 102
Neural tube birth defects, 17
New beverage introductions, 96
Nitrogen compounds, 31, 43, 51
No Action Indicated (NAI), 113
Nonbovine milk-related beverages, 27
Nonthermal juice processes, comparison of, 89
Nutraceutical(s)
   beverage, quintessential, 26
   FOSHU-approved, 19
   lipid class of, 51
Nutraceuticals, beverages as delivery systems for, 15–72
   biochemical, physiological, and molecular actions of nutraceuticals, 53–59
   classes of nutraceuticals, 45–53
   defining nutraceuticals/functional foods, 18–19
   future considerations, 59–60
   liquid foods, 19–44
Nutrients, immune enhancing, 57

O

OAI, *see* Official Action Indicated
Obesity, as worldwide health problem, 45
Official Action Indicated (OAI), 113
Olestra, 29, 46
Oligosaccharides, 20
Omega-3 fatty acids, 14, 56, 57
Oncogenes, 56
Oranges, inactivation of *E. coli* on surface of, 78
Organic growers, 196
Organic philosophy, 193
Organic products, USDA seal for, 200
Osteoporosis, 98
   calcium and, 16, 50
   prevention of, 53
Ox-Bar system, 208
Oxidation, 206
Oxygen scavenger, 206, 208
Ozonated water treatment, 139

P

Packaging
  active, 205–217
    antimicrobial polymers, 210–215
    bio-based materials for packaging, 215
    food labeling, 207
    oxygen scavengers/antioxidants, 207–210
    taint removers, 215–216
  antimicrobial, 212
  aseptic, 138
  materials
    starch-based polymer, 215
    using natural antimicrobials, 214
Paecilomyces, 101
Palantinose, 22
Papaya, virus-resistant, 10
Pasteurization, 79, 80, 135
  mandatory, 81, 137
  use of microwaves to heat food for, 87
  as organic process, 198
Pathogenic yeasts, 103
Pathogen reduction, 151
PCR, *see* Polymerase chain reaction
Pectin methylesterase (PME), 85
PEF, *see* Pulsed electric field
Penicillium, 88, 101
Peppermint oil, 49
Personal protective equipment (PPE), 186
Pest control measures, 187, 199
Pesticides, 4, 130, 182
Pharmafood, 24
Phenolics, 19, 24, 25, 32, 54
  antioxidant potency of, 28
  compounds, 51
  measurement of, 34
Phenylacetic acids, 32
Phenylpropenes, 32
Pichia membranaefaciens, 101
Pinene, 31
PLA, *see* Polylactate
Plant
  lignans, 52
  stanols, 29
  sterols, 21
Plastics, oxygen scavenger, 208
Platelet aggregation, 57
PME, *see* Pectin methylesterase
P450 mixed-function oxidase enzymes, 59
Polydextrose, 18, 20, 45
Polylactate (PLA), 215
Polymerase chain reaction (PCR), 12
Polymers
  antimicrobial, 210

  food acids added to heat-extruded, 216
Potassium sorbate, *E. coli* and, 78
Potato(es)
  solanine, 11
  virus-resistant, 10
Power Water, 27
PPE, *see* Personal protective equipment
Prebiotics, 43, 46
Preservatives, 78
Proanthocyanidins, 55
Probiotics, 27, 33, 42, 52
Produce
  foodborne outbreaks associated with, 3
  wash solution, 76
Product labeling, difficulties with, 12
Propionic acid bacterium, 21
Prostate cancer prevention, 51
Proteins, denatured, 12
Protozoa, 103
Prune juice, contaminated imported, 129
Psyllium, 20, 21, 58
Pulsed electric field (PEF), 79–81, 82, 83, 85

Q

QA, *see* Quality assurance
Quality
  assurance (QA), 189
  -control operations, 181
Quaternary ammonium salts, 211
Quercetin, 25
Quinones, 31

R

Radiation breeding, 10
RDAs, *see* Recommended dietary allowances
Reactive oxygen species (ROS), 56
Ready-to-eat (RTE) foods, 165
Recalled fruit drinks, 129
Recommended dietary allowances (RDAs), 17
Red Alert®, 41
Red Bull®, 41
Red Devil®, 41
Resistant starches, 31
Restriction fragment length polymorphism, 126
Resveratrol, 25, 42, 52
Rhizopus stolonifer, 88
Ribonucleic acid (RNA), 26
RNA, *see* Ribonucleic acid
ROS, *see* Reactive oxygen species
RTE foods, *see* Ready-to-eat foods

Index

S

Saccharin, 29, 45
Saccharobacter, 99
Saccharomyces
  bailii, 101
  bayanus, 101
  boulardii, 34
  cerevisiae, 82, 83
Salatrim, 29, 46
Salmonella, 74
  agona, 76
  anatum, 104
  chester, 76, 106
  enteritidis, 76, 104, 109, 127
  gaminara, 76, 104
  hartford, 104
  javiana, 106
  laboratory-confirmed cases of, 130
  manhattan, 127
  muenchen, 104, 108
  poona, 107
  runislaw, 104
  typhimurium, 76, 86, 84, 106
Sanitation
  activities, verification of, 189
  crew, members of, 177
  key elements in, 175–176
  schedule, master, 188
Sanitation, essential elements of, 175–192
  case studies, 190–192
  regulatory components of sanitation program, 177–184
    bottled water, 182–183
    GMPs, 179–182
    juice regulations, 183–184
    other regulations, 184
  sanitation needs, 176–177
  tools used in fulfilling sanitation program needs, 184–189
    cleaning and sanitizing, 184–186
    master sanitation schedule, 188–189
    other sanitation elements, 187–188
    procedures, 186
  verification of program efficacy, 189–190
Sanitation Standard Operating Procedures (SSOPs), 117, 145, 149, 160, 183
Saponins, 31
Sardine peptide, 21
Saw palmetto, 29, 46, 49
Schizosaccharomyces pombe, 101
Scientific Certification Systems (SCS), 6
SCS, *see* Scientific Certification Systems
Secoisolaricirasinol, 33
Self-heating cans, 206

Shelf life issues, 192
Shigella, 106, 127
Signaling, 55
Silent Spring, 194
Small round structured viruses (SRSVs), 103–104
Soda(s)
  caffeine in, 40–41
  definition of, 37
Soft drinks, 37, 40–41
Soil erosion, 10
Solanine, 11
Solstis®, 41
SOPs, *see* Standard Operating Procedures
Sorbitol, 29
Soy
  genistein in, 42
  milk, 27, 51
  -oligosaccharide, 20
  protein, 21, 25
Soybeans, herbicide-tolerant, 10
Sphingomyelins, 31
Spirits, 24
Spoilage bacteria in fruit juices and beverages, alternative processing technologies for, 73–93
  control of microbial contamination, 74–81
    pasteurization, 81
    preservatives, 78–81
    preventive measures in orchard, 74
    washing, 74–78
  irradiation, 87–88
  nonthermal alternative processing technologies, 81–87
    high pressure processing, 82–85
    pulsed electric field, 85–86
    ultraviolet light, 86–87
Spore germination, 85
Sport beverages, 38, 39
Squash, virus-resistant, 10
SRSVs, *see* Small round structured viruses
SSOPs, *see* Sanitation Standard Operating Procedures
Standard Operating Procedures (SOPs), 159–160
Starches, resistant, 31
StarLink™ incident, 13
Sterilization, electron beam, 139
Steroid glycosides, 31
Stilbenes, 32
Stimulator beverages, 40
St. John's wort, 29, 38, 46, 49
Stomach cancer, 55
Streptococcus salivarius, 20
Substantial equivalence, 11
Sucralose, 29, 45

Sugar(s)
  alcohols, 29
  fermentable, 73
  substitutes, 29, 38
Sulfites, 129
Sulfur compounds, 34, 42, 53
Sulfurofane, 14

T

Taint removers, 215
Talaromyces, 101
Taurine, 26, 40
Tea, 28
  CHD and, 28
  nutraceutical compounds found in, 28
  phenolics in, 35
  polyphenol, 22
Terpenes, 31
Tetraterpenes, 31, 36, 42
Third world
  perception of GAPs by, 3
  produce bought from, 4
Thirst quencher, water as, 39
Time–temperature sensors, unique use of, 206
Toilet facilities, maintenance of, 172
Tomato, lycopene in, 14, 41
Torulaspora delbruckii, 101
Torulopsis holmii, 101
Toxicants, natural, 11, 13
Toxic compounds, labeling, storage, and use of, 172
Transposon mutagenesis, 10
Triglycerides, 26, 29, 45
Triterpenes, 31

U

UL, *see* Upper Limits
United Fresh Fruit and Vegetable Association, 5
Unpasteurized cider, 81
Upper Limits (UL), 24, 44
Urinary tract infections (UTIs), 37, 55
USDA, *see* U.S. Department of Agriculture
U.S. Department of Agriculture (USDA), 10–11, 112, 195
  National Organic Program, 201, 202, 203
  organic regulations, 195
  seal for organic products, 200
U.S. Food and Drug Administration, *see* Food and Drug Administration
UTIs, *see* Urinary tract infections

V

VAI, *see* Voluntary Action Indicated
Valerian root, 29, 49
Vialize®, 41
Vibrio cholerae, 126
Viruses, 10, 103
Vitamin E, 14
Vitamin-fortified organic beverage, 197
Vitamin Water, 27
V8 juice, nutraceuticals provided in, 37
Voluntary Action Indicated (VAI), 113

W

Wastewater, BOD of, 184
Water
  bottled, 182
  as essential nutrient, 26
  Fitness, 27
  Fortified, 27
  Power, 27
  Vitamin, 27
Wellness beverages, 41
Wheat
  albumin, 22
  bran, 20, 43
Wine, 24
  anthocyanidins in, 52
  consumption, CHD and, 25
  phenolics in, 19
  resveratrol in, 42

X

Xanthones, 32
Xanthophylls, 25, 36, 54
Xylitol, 29
Xylo-oligosaccharide, 20

Y

Yeast(s), 101
  contamination, 110
  pathogenic, 103
Yersinia, 106, 127
Yogurt(s), 27
  drink-type, 20
  starter culture organisms in, 19

Z

Zeaxanthin, 31, 36, 37, 51, 42
Zygosaccharomyces microellipsoides, 101
Zymobacter, 99
Zymomonas, 99